International Series in Operations Research & Management Science

Founding Editor
Frederick S. Hillier

Volume 375

The book series **International Series in Operations Research and Management Science** encompasses the various areas of operations research and management science. Both theoretical and applied books are included. It describes current advances anywhere in the world that are at the cutting edge of the field. The series is aimed especially at researchers, advanced graduate students, and sophisticated practitioners.

The series features three types of books:

• Advanced expository books that extend and unify our understanding of particular areas.

• Research monographs that make substantial contributions to knowledge.

• Handbooks that define the new state of the art in particular areas. Each handbook will be edited by a leading authority in the area who will organize a team of experts on various aspects of the topic to write individual chapters. A handbook may emphasize expository surveys or completely new advances (either research or applications) or a combination of both.

The series emphasizes the following four areas:

Mathematical Programming: Including linear programming, integer programming, nonlinear programming, interior point methods, game theory, network optimization models, combinatorics, equilibrium programming, complementarity theory, multiobjective optimization, dynamic programming, stochastic programming, complexity theory, etc.

Applied Probability: Including queuing theory, simulation, renewal theory, Brownian motion and diffusion processes, decision analysis, Markov decision processes, reliability theory, forecasting, other stochastic processes motivated by applications, etc.

Production and Operations Management: Including inventory theory, production scheduling, capacity planning, facility location, supply chain management, distribution systems, materials requirements planning, just-in-time systems, flexible manufacturing systems, design of production lines, logistical planning, strategic issues, etc.

Applications of Operations Research and Management Science: Including telecommunications, health care, capital budgeting and finance, economics, marketing, public policy, military operations research, humanitarian relief and disaster mitigation, service operations, transportation systems, etc.

This book series is indexed in Scopus.

Ghaith Rabadi • Bulent Soykan
Editors

Optimizing Supply Chains Through Digital Twins

Navigating the Modern Landscape

Springer

Editors
Ghaith Rabadi
School of Modeling, Simulation and Training
University of Central Florida
Orlando, FL, USA

Bulent Soykan
Institute for Simulation and Training
University of Central Florida
Orlando, FL, USA

ISSN 0884-8289 ISSN 2214-7934 (electronic)
International Series in Operations Research & Management Science
ISBN 978-3-032-08146-9 ISBN 978-3-032-08147-6 (eBook)
https://doi.org/10.1007/978-3-032-08147-6

This Springer imprint is published by the registered company Springer Nature Switzerland AG
The registered company address is: Gewerbestrasse 11, 6330 Cham, Switzerland

Contents

Contributors

Raid Al-Aomar German-Jordanian University, Amman, Jordan

Ola G. Batarseh Dassault Systemes, Chennai, Tamil Nadu, India
University of Massachusetts Lowell, Lowell, MA, USA

Fabian Behrendt Magdeburg-Stendal University of Applied Science, Magdeburg, Germany

Rajinder Bhandal University of Leeds, Leeds, UK

Bahar Biller SAS Institute, Cary, NC, USA

Stephan R. Biller Purdue University, West Lafayette, IN, USA

Grace Bochenek School of Modeling, Simulation and Training, University of Central Florida, Orlando, FL, USA

Canan Gunes Corlu Metropolitan College, Boston University, Boston, MA, USA

Sean Jefferson Johnson & Johnson, Santa Clara, CA, USA

Masoud Kahalimoghadam Faculty of Engineering and IT, The University of Melbourne, Melbourne, Australia

Michail Katsigiannis Department of Industrial and Systems Engineering, Auburn University, Auburn, AL, USA

Gulsah Hancerliogullari Koksalmis Department of Industrial Engineering and Management Systems, University of Central Florida, Orlando, FL, USA
Faculty of Management, Department of Industrial Engineering, Istanbul Technical University, Istanbul, Türkiye

Emrah Koksalmis Department of Industrial Engineering, Turkish Air Force Academy, National Defense University, Istanbul, Türkiye

Sean Mondesire School of Modeling, Simulation and Training, University of Central Florida, Orlando, FL, USA

Rami Musa Johnson & Johnson, West Chester, PA, USA

Konstantinos Mykoniatis Department of Industrial and Systems Engineering, Auburn University, Auburn, AL, USA

Ghaith Rabadi School of Modeling, Simulation and Training, University of Central Florida, Orlando, FL, USA

Soheil Sabri School of Modeling, Simulation and Training, University of Central Florida, Orlando, FL, USA

Pieter van Schalkwyk XMPro, Dallas, TX, USA

Bulent Soykan Institute for Simulation and Training, University of Central Florida, Orlando, FL, USA

Russell G. Thompson Faculty of Engineering and IT, The University of Melbourne, Melbourne, Australia

Tim Wollert Magdeburg-Stendal University of Applied Science, Magdeburg, Germany

Lele Zhang School of Mathematics and Statistics, The University of Melbourne, Melbourne, Australia

Part I
Introduction and Foundations

Chapter 1
Introduction

Bulent Soykan and Ghaith Rabadi

The landscape of global supply chains has never been more complex or volatile. Decades of globalization created intricate, geographically dispersed networks offering efficiency but also introducing inherent fragility. In recent years, these networks have faced unprecedented pressure from a confluence of forces: disruptive global events like pandemics, escalating geopolitical tensions, rapid shifts in consumer demand, increasing calls for sustainability, and the relentless pace of technological change. Traditional approaches to supply chain management, often relying on historical data, periodic planning cycles, and siloed operations, are increasingly proving insufficient to navigate this turbulent environment. The need for enhanced visibility, predictive foresight, operational agility, and robust resilience is no longer just a competitive advantage—it is a necessity for survival and success.

In response to these profound challenges, a powerful paradigm shift is underway, driven by digital transformation. At the forefront of this shift is the concept of the Digital Twin (DT). No longer confined to the realms of product design or manufacturing assets, the Digital Twin concept is being increasingly applied to entire complex systems, including the end-to-end supply chain. At its core, a supply chain Digital Twin is a dynamic virtual representation of the physical supply chain—its assets, processes, relationships, and behaviors. Crucially, it is synchronized with its physical counterpart through real-time data streams and incorporates a feedback loop, allowing the virtual model to not only mirror reality but also to inform and influence physical operations.

B. Soykan (✉)
Institute for Simulation and Training, University of Central Florida, Orlando, FL, USA
e-mail: Bulent.Soykan@ucf.edu

G. Rabadi
School of Modeling, Simulation and Training, University of Central Florida, Orlando, FL, USA
e-mail: Ghaith.Rabadi@ucf.edu

G. Rabadi, B. Soykan (eds.), *Optimizing Supply Chains Through Digital Twins*,
International Series in Operations Research & Management Science 375,
https://doi.org/10.1007/978-3-032-08147-6_1

This ability to create a living, evolving digital replica offers transformative potential. By integrating vast amounts of data from diverse sources—IoT sensors tracking shipments, enterprise resource planning (ERP) systems managing orders, warehouse management systems (WMS) detailing inventory, transportation management systems (TMS) monitoring logistics, and external feeds providing market or weather data—Digital Twins provide unprecedented visibility across the network. But they go far beyond mere visualization. Embedded analytics, simulation capabilities, and increasingly, artificial intelligence (AI) and machine learning (ML), allow these Digital Twins to perform sophisticated analysis, enabling prediction of future states, simulation of "what-if" and "what-next" scenarios, and prescription of optimal actions. From anticipating demand fluctuations and identifying potential bottlenecks before they occur, to optimizing inventory levels across multiple echelons and stress-testing resilience against disruptions, Digital Twins empower organizations to move from reactive problem-solving to proactive, data-driven optimization and risk management.

Realizing this potential, however, requires more than just implementing a single piece of software. The power of a supply chain Digital Twin lies in the synergistic convergence of multiple enabling technologies. The Internet of Things (IoT) provides the sensory input, capturing real-time data from the physical world. Cloud Computing offers the scalable infrastructure needed to store and process immense datasets and complex models. Advanced Data Analytics, including AI and ML, furnishes the intelligence to extract insights, make predictions, and drive automation. Simulation Modeling provides the virtual testbed for experimentation and optimization. Blockchain offers potential for enhancing trust and transparency in multi-party transactions. Furthermore, robust architectural frameworks are essential to structure these complex ecosystems, ensuring data flows correctly and components integrate seamlessly. Understanding this technological backbone and the methodologies for architecting and leveraging these integrated systems is critical for successful Digital Twin implementation.

1.1 Purpose and Scope of this Book

"Optimizing Supply Chains Through Digital Twins—Navigating the Modern Landscape" aims to provide a comprehensive and insightful guide to this rapidly evolving field. Recognizing the need for a resource that bridges foundational concepts with practical applications and future considerations, this book brings together expertise spanning the core components of supply chain Digital Twins. Our objective is to equip supply chain professionals, managers, consultants, researchers, and students with the knowledge needed to understand, evaluate, and begin leveraging Digital Twin technology to optimize their operations and navigate the complexities of the modern supply chain environment.

This book investigates:

1. The fundamental concepts and value proposition of Digital Twins in supply chain management.
2. The key enabling technologies that form the digital backbone.
3. Methodologies and frameworks for designing, developing, and implementing DTs, including simulation, agent-based modeling, and architectural approaches.
4. Critical technical considerations, such as managing and quantifying uncertainty in simulations.
5. Diverse applications and case studies, from control towers, automotive manufacturing, to hyperconnected city logistics and semiconductor supply chains.
6. The integration of advanced capabilities like AI and ML, and the associated research gaps and challenges that must be addressed.

1.2 Structure of the Book

To guide the reader through this multifaceted topic, the book is organized into five logical parts:

Part 1: Introduction and Foundations: We begin by establishing the context (Chap. 1—Introduction—Charting the Course for Supply Chain Optimization with Digital Twins), reviewing the current state of research through bibliometric analysis (Chap. 2—Digital Twin in Supply Chain Management: A Comprehensive Bibliometric Review), defining the core concepts of Digital Twins and their relationship with simulation (Chap. 3—Harnessing the Power of Simulation in Supply Chain Digital Twins), and introducing the pivotal application of Control Towers as a manifestation of DT principles (Chap. 4—Supply Chain Digital Twin Application: Control Tower).

Part 2: Enabling Technologies and Architecture: This part focuses on the technological underpinnings and architectural design, detailing the crucial roles of enabling technologies (Chap. 5—The Digital Backbone of Modern Supply Chains: Enabling Technologies for Digital Twins), presenting methodologies for formally architecting the supply chain ecosystem (Chap. 6—Modeling Supply Chain Ecosystem Architecture to Support Digital Twins), and illustrating a specific, detailed architecture for the complex semiconductor industry (Chap. 7—A Supply Chain Digital Twin Architecture for Semiconductor Industry).

Part 3: Methodologies and Frameworks: In this part, we explore specific approaches for building and utilizing DTs, including detailed frameworks for simulation-based twins and Value Stream Digital Twins (Chap. 8—A Framework for Supply Chain Simulation and Digital Twinning), the application of agent-based modeling for optimization and immunity (Chap. 9—"Navigating the Modern Landslide": The Agent-Based View Towards Supply Chain Immunity and Digital Twin Optimization—A Use Case with XMPro), and techniques for addressing the critical issue of uncertainty quantification in DT simulations (Chap. 10—Uncertainty Quantification in Digital Twin Simulations).

Part 4: Specific Applications and Case Studies: This part showcases the practical application of Digital Twins in specific domains, offering in-depth case studies in automotive assembly lines (Chap. 11—Digitalizing the Automotive Assembly Line: A Case Study for Enhancing Supply Chain Efficiency with Digital Twin Technology) and the complex environment of hyperconnected city logistics (Chap. 12—Digital Twins for Hyperconnected City Logistics).

Part 5: Advanced Integration and Future Directions: Finally, we look towards the cutting edge and the future, exploring the powerful synergy of integrating AI and Machine Learning with Digital Twins, while also critically examining the current research gaps and foundational challenges that need to be overcome (Chap. 13—Integrating AI and ML in Supply Chain Digital Twins: Bridging Potential and Foundational Research Gaps).

By navigating through these interconnected themes, readers will gain a holistic understanding of how Digital Twins are reshaping supply chain management. This book serves as both a primer for those new to the concept and a valuable reference for practitioners seeking to implement or enhance Digital Twin capabilities within their organizations. The journey towards truly optimized, resilient, and intelligent supply chains is complex, but Digital Twins provide an indispensable map and compass for navigating the modern landscape.

Chapter 2
Digital Twin in Supply Chain Management: A Comprehensive Bibliometric Review

Gulsah Hancerliogullari Koksalmis and Emrah Koksalmis

Abstract The incorporation of Digital Twin (DT) technology into Supply Chain Management (SCM) represents a paradigm shift, offering enhanced efficiency, resilience, and strategic insights for managing global supply networks. This study provides an extensive bibliometric analysis, highlighting key trends, research themes, influential authors, and leading institutions in this dynamic domain. The analysis utilizes a dataset of scholarly publications from 2014 to 2025, sourced from the Scopus database, a reliable repository for academic research. By employing bibliometric techniques such as co-citation analysis, bibliographic coupling, collaboration mapping, and keyword co-occurrence analysis, the study identifies pivotal contributors, prominent institutions, and impactful nations based on citation metrics. The findings offer a comprehensive overview of DT applications in SCM, equipping researchers, practitioners, and policymakers with actionable insights to leverage this technology for optimizing supply chain performance.

Keywords Bibliometric analysis · Supply chain management · Digital twin · Scopus · VOSviewer

G. H. Koksalmis (✉)
Department of Industrial Engineering and Management Systems, University of Central Florida, Orlando, FL, USA

Faculty of Management, Department of Industrial Engineering, Istanbul Technical University, Istanbul, Türkiye
e-mail: gulsah.hancerliogullarikoksalmis@ucf.edu

E. Koksalmis
Department of Industrial Engineering, Turkish Air Force Academy, National Defense University, Istanbul, Türkiye

G. Rabadi, B. Soykan (eds.), *Optimizing Supply Chains Through Digital Twins*, International Series in Operations Research & Management Science 375,
https://doi.org/10.1007/978-3-032-08147-6_2

2.1 Introduction

The concept of a "digital twin" emerges from the convergence of physical systems with their virtual counterparts, creating dynamic, data-driven representations of real-world entities. First conceptualized by Michael Grieves in 2002 within the context of product lifecycle management, digital twins have evolved into a transformative technology with diverse applications, including manufacturing, healthcare, and urban development. These systems consist of three integral components: the physical entity, its digital model, and a data connection that ensures real-time synchronization between the two (Tao et al., 2019).

Over the past decade, rapid advancements in the Internet of Things (IoT), artificial intelligence (AI), and data analytics have accelerated the adoption of digital twin technology, positioning it as a cornerstone of operational excellence across industries. Within the realm of supply chain management, digital twins facilitate real-time monitoring, predictive analytics, and process optimization by modeling virtual replicas of supply chain operations, facilities, and assets (Ivanov et al., 2021). These models empower stakeholders to simulate scenarios, pinpoint inefficiencies, and implement proactive measures, making them indispensable for navigating the complexities and uncertainties of modern supply chains.

The disruptions caused by the COVID-19 pandemic underscored the critical need for resilient and adaptive supply chain systems. Challenges such as lockdowns, workforce shortages, and fluctuating demand highlighted the necessity for technologies capable of delivering real-time insights and dynamic reconfiguration. Digital twins emerged as a pivotal solution, providing capabilities that surpass traditional supply chain management approaches.

Existing research has largely focused on specific applications or narrow areas, such as logistics and inventory management. This paper addresses these gaps by offering a holistic examination of digital twin technology within SCM. It analyzes the current state, explores potential applications, and outlines future research directions, aiming to contribute valuable insights to the growing body of literature and support practitioners in creating smarter, more resilient supply chains.

Research Objectives

This study aims to investigate the bibliometric characteristics of digital twin applications in supply chain management, analyzing publication trends and citation patterns up to 2025. The specific goals include:

- Examining the annual output of publications and citations for the most-cited articles.
- Identifying the most active countries, institutions, and authors contributing to this field.
- Visualizing research content through bibliometric mapping and thematic analysis to reveal collaboration patterns.

To achieve these objectives, the following research questions are addressed:

- Which countries, authors, and institutions lead in publication output?
- How are publications categorized by document type and academic discipline?
- Who are the most influential researchers in this domain?
- What patterns of collaboration exist among researchers?
- Which nations and affiliations exert the greatest influence in this field?

The paper is structured as follows: Section 2.2 describes the research methodology, including data collection and analysis techniques. Section 2.3 presents the results, focusing on trends, relationships, and insights derived from bibliometric coupling network analysis. Section 2.4 concludes with a discussion of the study's contributions and implications.

2.2 Methodology

2.2.1 Research Design

This research employs bibliometric analysis to investigate scientific publications on digital twins in supply chain management. It adopts a bibliographic coupling network approach to uncover trends and relationships within the field.

2.2.2 Data Collection

The dataset is sourced from the Scopus database, renowned for its extensive academic coverage. As of December 18, 2024, 5550 relevant publications are identified using the search terms "supply chain management" (All fields) AND "digital twin" (All fields). The data collection process is depicted in Fig. 2.1, which provides a flowchart of the methodology.

For the bibliometric analysis, this study employs **VOSviewer** and **Microsoft Excel.** VOSviewer is instrumental in analyzing and visualizing complex bibliographic datasets, enabling the identification of networks and clusters within the research landscape. In contrast, Microsoft Excel is utilized for generating summary tables and creating charts to present the data in a concise and accessible format.

2.3 Data Analysis and Results

2.3.1 Bibliometric Trends

Figure 2.2 depicts the annual trends in publication output from 2014 to 2025, highlighting a substantial rise in research activity over recent years. Since 2018,

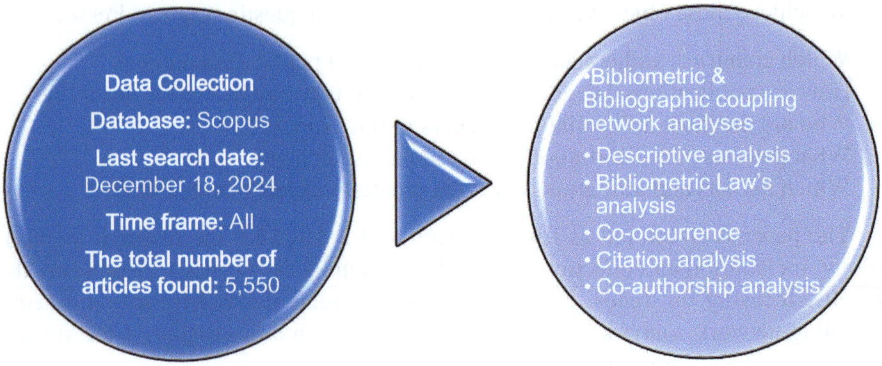

Fig. 2.1 Flow chart for the data collection process

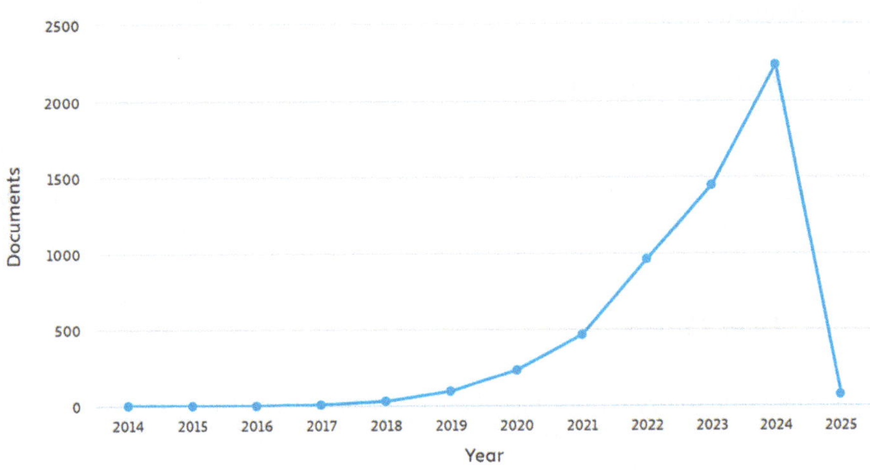

Fig. 2.2 Publication trends according to years

there has been a steady growth in the number of publications in this domain. The observed drop in 2025 can be attributed to the timing of data collection, which is conducted in late 2024, leading to incomplete reporting for the subsequent year.

Figure 2.3 presents the citation trends for research on digital twin technology in supply chain management from 2014 to 2024. The data reveals a consistent rise in the total number of citations over the years. This upward trajectory underscores the increasing recognition and influence of this research area in the academic community, highlighting its expanding importance and credibility within the field.

Figure 2.4 displays the top five journals publishing research on digital twin technology in supply chain management. *Sustainability* leads the list with 287 publications, followed by *International Journal of Production Research* with 125, *Journal of Cleaner Production* with 102, *IEEE Access* with 97, and *Applied Sciences*

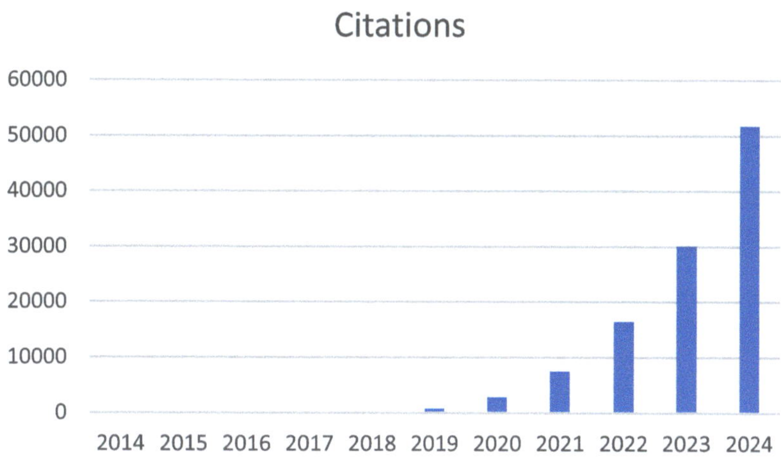

Fig. 2.3 Number of citations over the years

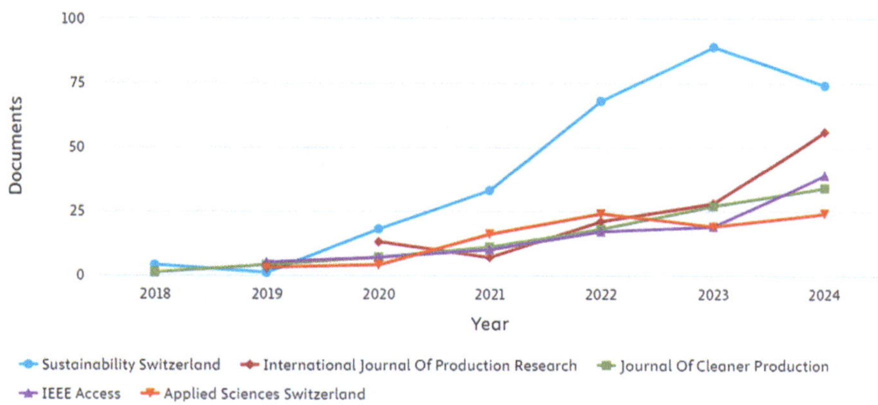

Fig. 2.4 Documents per year by source

with 90 publications. These journals represent the primary outlets for advancing research in this field.

Figure 2.5 highlights the top ten countries contributing to research on digital twin technology in supply chain management. China leads with 1023 publications, followed by India with 893, the United Kingdom with 588, and the United States with 529 publications. These countries are the primary drivers of scholarly activity in this field.

Figure 2.6 categorizes the types of publications related to digital twin technology in supply chain management. The majority of publications are classified as *Articles* (3186), followed by *Conference Papers* (849), *Reviews* (807), *Book Chapters* (521), *Books* (128), *Editorials* (30), *Retracted* (10), *Conference Reviews* (9), *Short Surveys*

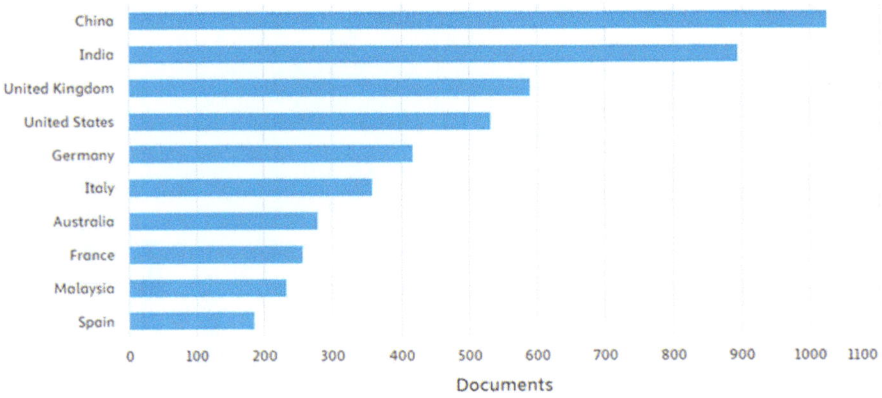

Fig. 2.5 Publication according to countries

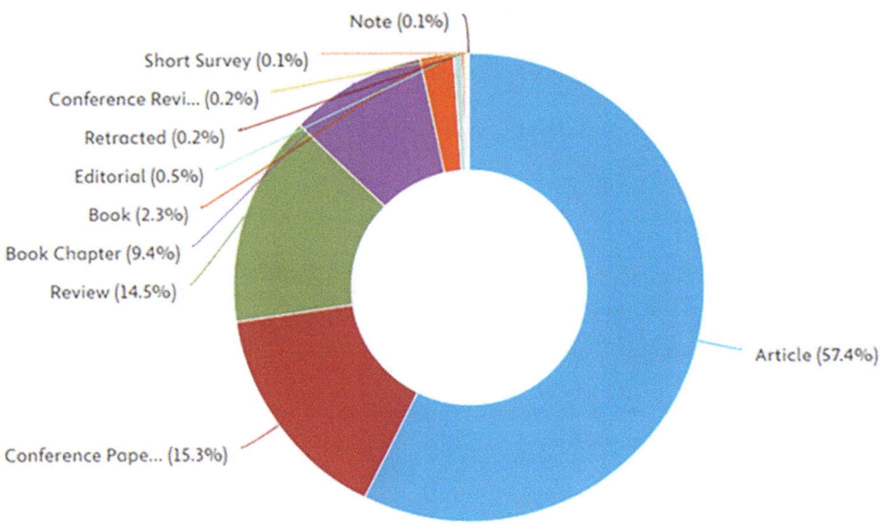

Fig. 2.6 Publication according to document type

(7), and *Notes* (3), among others. This distribution reflects the diverse forms of scholarly output in this growing research area.

Figure 2.7 illustrates the various subject areas covered by the journal, including *Engineering*, *Computer Science*, *Business, Management, and Accounting*, *Decision Sciences*, *Social Sciences*, *Environmental Science*, *Energy*, *Mathematics*, *Economics, Econometrics and Finance*, *Materials Science*, and others. This broad range of disciplines highlights the interdisciplinary nature of research on digital twin technology in supply chain management.

Figure 2.8 presents the most prolific affiliations contributing to research on digital twin technology in supply chain management. The Hong Kong Polytechnic

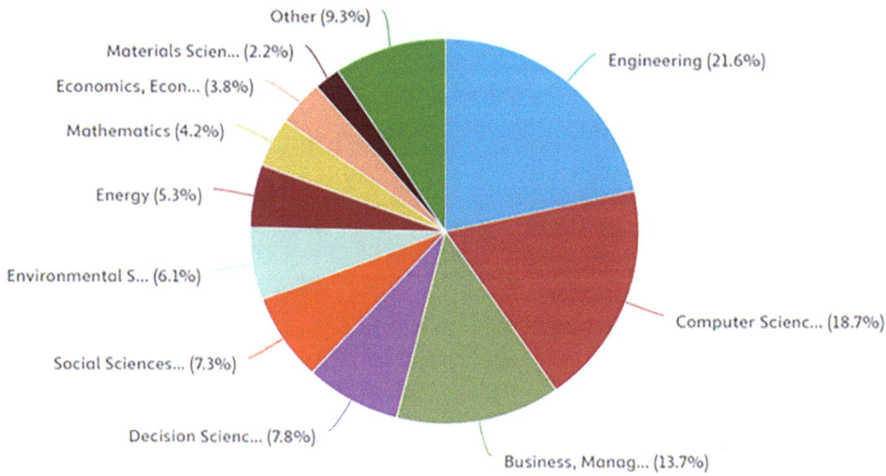

Fig. 2.7 Publication according to subject type

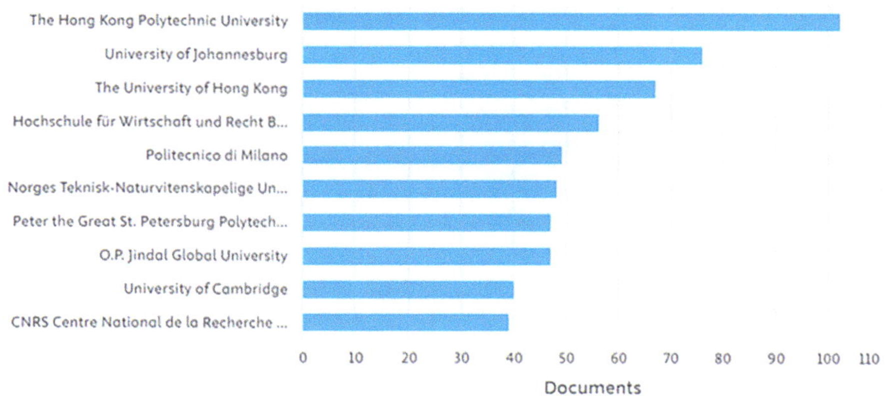

Fig. 2.8 Publication according to affiliations

University leads with 102 publications, then the University of Johannesburg with 76, the University of Hong Kong with 67, and Hochschule für Wirtschaft und Recht Berlin with 56 publications. These institutions are key contributors to the advancement of research in this field.

Figure 2.9 highlights the top ten authors who have made significant contributions to the research on digital twin technology in supply chain management. Ivanov, D. leads with 57 publications, followed by Dolgui, A. with 26, Javaid, M. with 26, Haleem, A. with 25, and Huang, G.Q. with 25 publications. These authors are among the most influential researchers in the field.

Figure 2.10 presents the top ten funding sponsors for research on digital twin technology in supply chain management. The National Natural Science Foundation

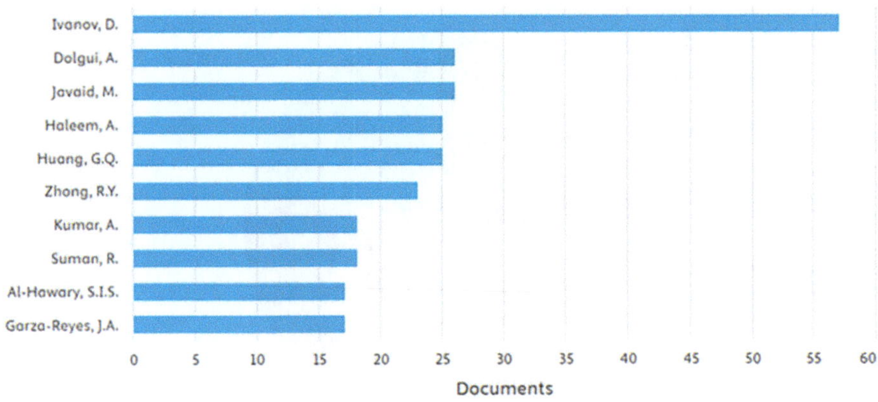

Fig. 2.9 Publication according to authors

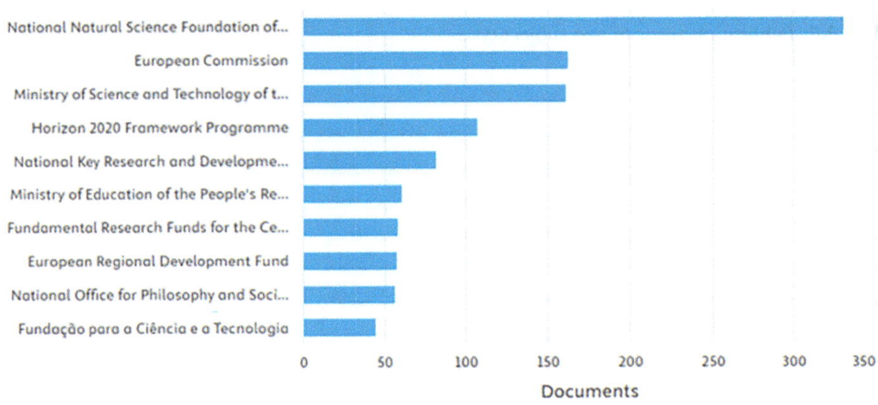

Fig. 2.10 Documents by funding sponsor

of China is the leading sponsor, supporting 330 publications, followed by the European Commission with 162, the Ministry of Science and Technology of the People's Republic of China with 161, the Horizon 2020 Framework Programme with 107, and the National Key Research and Development Program of China with 81 publications. These organizations play a crucial role in funding and advancing research in this area.

A co-occurrence analysis is done on "author keywords" that appeared more than five times, as shown in Fig. 2.11. Out of 11,105 keywords, 691 met the threshold, resulting in 14 distinct clusters. The largest cluster, labeled "*Industry 4.0*," contains 109 keywords. The second-largest cluster, "*Blockchain*," includes 77 keywords, while the third cluster, "*Digital Twin*," consists of 66 keywords. This analysis highlights the key themes and their interconnections within the field.

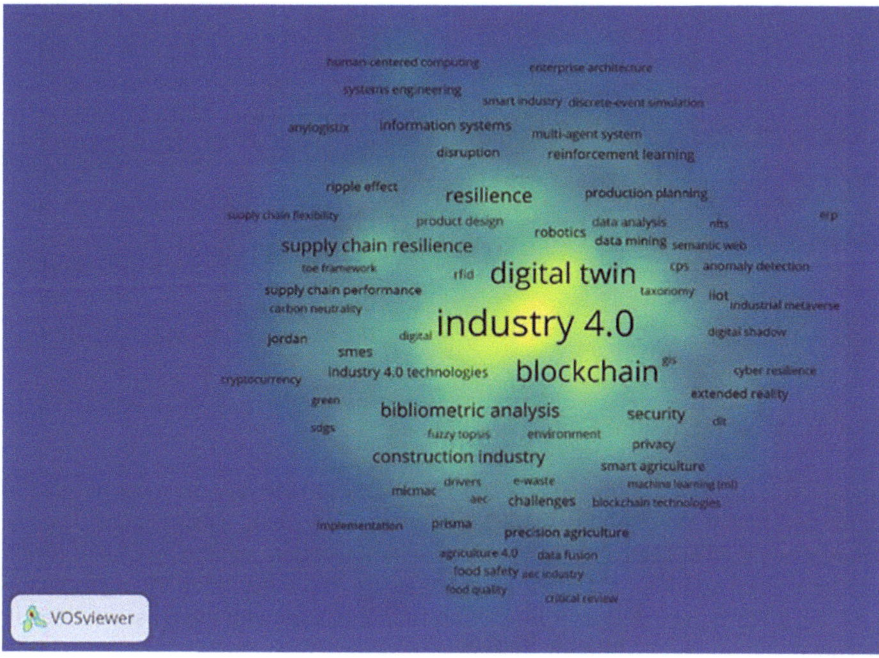

Fig. 2.11 Analysis of co-occurring of author-keywords

Table 2.1 Leading contributors based on citation

Contributor	Records	Citations
M. Ghobakhloo	3	2366
D. Ivanov	12	1987
M. Javaid et al.	6	928
D. Ivanov, A. Dolgoui	3	840
M. Soori et al.	3	470
T.D. Moshood	3	335
A. Dolgui, D. Ivanov	3	217
M. Perno et al.	3	197
I. Ante	3	166
K. Govindan	3	164

2.3.1.1 Most Influential Researchers

To identify the leading contributors to the field of digital twin technology in supply chain management, a citation analysis is conducted on 5390 authors. Table 2.1 ranks the top 10 authors based on their citation counts. According to this analysis, M. Ghobakhloo ranks first, followed by D. Ivanov in second place, and M. Javaid et al. in third place. These authors have made the most significant contributions to the field in terms of citations.

Table 2.2 Top countries based on citation

Country	Records	Citations
China	1023	21,305
India	894	17,937
United Kingdom	588	15,775
United States	531	14,714
France	255	12,383
Germany	416	12,378
Italy	356	10,221
Australia	277	8875
Brazil	160	7782
Hong Kong	163	7079

2.3.1.2 Most Influential Countries

A citation analysis is conducted across 159 countries to identify the most influential nations in the field of digital twin technology in supply chain management. The top countries, ranked in Table 2.2 based on their citation counts, are led by China, which has the highest citation total. Following China are India, the United Kingdom, and the United States, which round out the top four countries with the most citations in this area.

2.3.1.3 Most Influential Affiliations

A citation analysis is performed on 12,665 affiliations in order to find out the leading institutions in the research area of digital twin technology in supply chain management. Table 2.3 presents the top 10 affiliations based on their citation scores. The Federal University of Rio Grande do Sul Organizational Engineering Group leads in citation counts, followed by the Jamia Millia Islamia Department of Mechanical Engineering, and the Govind Ballabh Pant University Department of Industrial & Production Engineering. These institutions are the most influential in terms of citations in this field.

2.3.1.4 Co-authorship Analysis

A co-authorship analysis is done for countries whose authors collaborate on digital twin technology in supply chain management, focusing on those with at least five publications. Out of the 159 countries, 86 met this criterion, leading to the identification of nine distinct clusters, as shown in Fig. 2.12. The analysis demonstrates that nations are connected within the network. The most leading nations, according to Total Link Strength (TLS), include China, India, the United Kingdom, the United States, France, Germany, Italy, Australia, Brazil, and Hong Kong. These countries stand out as key contributors to the global research network.

Table 2.3 Most influential affiliations based on citation

Organization	Records	Citations
Federal University of Rio Grande do Sul, Organizational Engineering Group	8	3684
Jamia Millia Islamia Department of Mechanical Engineering	17	1490
Govind Ballabh Pant University Department of Industrial & Production Engineering	11	1200
Berlin School of Economics and Law, Supply Chain and Operations Management	5	1037
Kaunas University of Technology, School of Economics and Business	8	886
Dr B R Ambedkar National Institute of Technology, Department of Industrial and Production Engineering	8	880
University of Hong Kong, Department of Industrial and Manufacturing Systems Engineering	20	819
Hong Kong Polytechnic University, Department of Industrial and Systems Engineering	7	639
Asia University Institute of Innovation and Circular Economy	9	601
EMPA Swiss Federal Laboratories for Materials Science and Technology	6	510

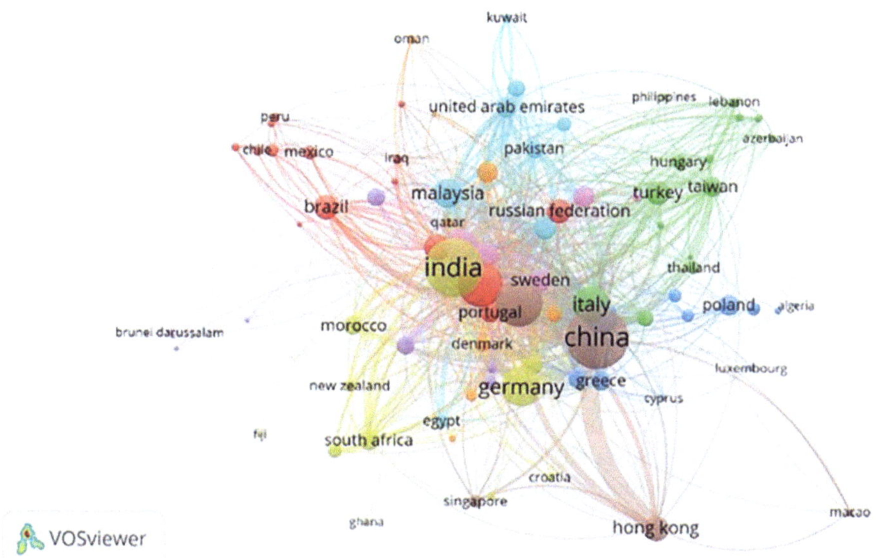

Fig. 2.12 Map of the co-authorship network

Similarly, a co-authorship analysis is done for affiliations with at least five publications. Out of 12,655 affiliations, 77 met this threshold, resulting in the identification of four clusters, as shown in Fig. 2.13. The top five affiliations based on citation scores are led by the Federal University of Rio Grande do Sul

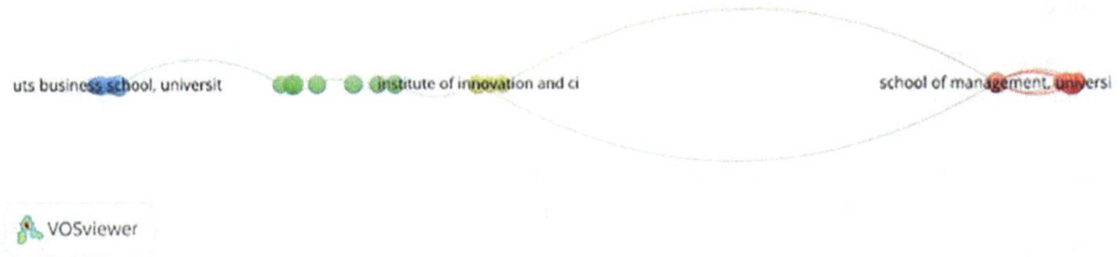

Fig. 2.13 Map of affiliations based on co-authorship connected clusters

Organizational Engineering Group, followed by Jamia Millia Islamia Department of Mechanical Engineering, Govind Ballabh Pant University Department of Industrial & Production Engineering, Berlin School of Economics and Law Supply Chain and Operations Management, and Kaunas University of Technology School of Economics and Business. These institutions represent the most influential contributors to the field of digital twin technology in supply chain management.

2.3.2 Bibliographic Coupling Network Analysis

Bibliographic coupling refers to the situation where two different publications cite the same source, indicating a similarity in their topics or other characteristics (Kurutkan & Orhan, 2018). This type of coupling occurs when two studies reference one or more common works, signifying that they rely on the same foundational material. Essentially, it can be understood as a connection between two documents that share overlapping bibliographies.

Figure 2.14 depicts the bibliographic coupling network for documents with at least 20 citations. Out of the 5550 documents analyzed, 997 met the criteria, with 640 being interconnected. This analysis reveals several distinct clusters, which include:

- Digital Twins and Reinforcement Learning in Supply Chains
- Digitalization for Sustainable and Circular Industrial Systems
- Digital Innovation and Sustainability
- Digital Twins and Blockchain
- Blockchain for Supply Chain
- Digital Transformation in Manufacturing and Supply Chains Post-COVID
- Resilient Manufacturing and Digital Transformation

Table 2.4 shows the ten most cited documents, along with their respective clusters and citation counts per year. The article "Industry 4.0 technologies: Implementation patterns in manufacturing companies" by Frank et al. (2019) leads with 1858 citations, making it the most cited document. Following closely, the works "Predicting the impacts of epidemic outbreaks on global supply chains: A simulation-based

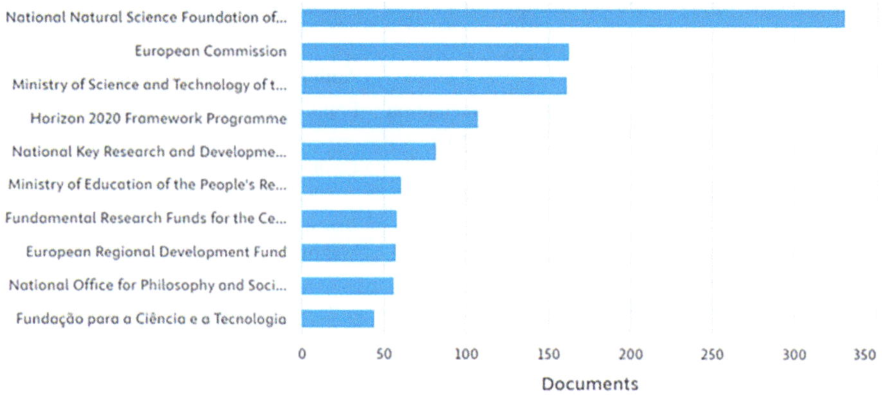

Fig. 2.14 Connected document bibliographic coupling network

analysis on the coronavirus outbreak (COVID-19/SARS-CoV-2) case" by Ivanov (2020), "Literature review of Industry 4.0 and related technologies" by Oztemel and Gursev (2020), and "Metaverse beyond the hype: Multidisciplinary perspectives on emerging challenges, opportunities, and agenda for research, practice and policy" by Dwivedi et al. (2022) are ranked third, fourth, and sixth, respectively, highlighting their recent relevance.

2.4 Conclusion

The concept of the "digital twin" has transformed industries by bridging physical systems with dynamic virtual counterparts, enabling real-time insights and optimization. Originating from Michael Grieves' 2002 work on product lifecycle management, digital twins have become integral to supply chain management, particularly in addressing disruptions like those caused by COVID-19. This study examines the bibliometric analysis of digital twin applications in supply chains, analyzing trends, key contributors, and collaboration networks. By offering a comprehensive overview, it aims to support the development of smarter, more resilient supply chains and guide future research in this evolving field.

To the best of our knowledge, no bibliometric analysis has been conducted on the application of digital twin technology in supply chain management. This research aims to address this gap by conducting a comprehensive bibliometric analysis of the field. By examining 5550 articles from the Scopus database, spanning from 2014 to 2025, this study offers valuable insights into the evolution of digital twin technology in supply chain management.

The initial phase of the analysis provides a quantitative review of the publication trends, including yearly output, countries of origin, languages, document types,

Table 2.4 The 10 most cited documents

Rank	Title				
Citation	Cluster	Author	Year	Total	
1	"Industry 4.0 technologies: Implementation patterns in manufacturing companies"	Frank et al.	2019	1858	2
2	"A systematic literature review of blockchain-based applications: Current status, classification and open issues"	Casino et al.	2019	1517	5
3	"Predicting the impacts of epidemic outbreaks on global supply chains: A simulation-based analysis on the coronavirus outbreak (COVID-19/SARS-CoV-2) case"	Ivanov	2020	1486	6
4	"The expected contribution of Industry 4.0 technologies for industrial performance"	Dalenogare et al.	2018	1433	2
5	"Literature review of Industry 4.0 and related technologies"	Oztemel and Gursev	2020	1366	4
6	"Metaverse beyond the hype: Multidisciplinary perspectives on emerging challenges, opportunities, and agenda for research, practice and policy"	Dwivedi et al.	2022	1232	3
7	"Industry 4.0, digitization, and opportunities for sustainability"	Ghobakhloo	2020	1202	7
8	"Industry 5.0: A survey on enabling technologies and potential applications"	Maddikunta et al.	2022	1012	2
9	"The future of manufacturing industry: a strategic roadmap toward Industry 4.0"	Ghobakhloo	2018	963	3
10	"Sustainable Industry 4.0 framework: A systematic literature review identifying the current trends and future perspectives"	Kamble et al.	2018	944	2

research domains, author affiliations, and frequently used keywords. This study also identifies the ten most cited articles in the field. To further enhance the understanding of the literature, a bibliographic network analysis is performed, revealing seven distinct clusters of articles. Each cluster is analyzed, with thematic labels assigned based on the content.

The analysis shows that China, India, the United Kingdom, and the United States are the leading contributors in terms of publication output, and these countries also have the highest citation frequencies. This trend suggests that countries with higher publication volumes tend to exert greater influence in the field. In terms of affiliations, Hong Kong Polytechnic University, the University of Johannesburg, and the University of Hong Kong lead in publication volume, while the most cited affiliations include the Federal University of Rio Grande do Sul, Jamia Millia Islamia, and Govind Ballabh Pant University. This finding indicates that a higher publication output does not always correlate with higher citation frequency.

Among the authors, Ivanov, D., Dolgui, A., and Javaid, M. have the highest publication counts. When ranked by citation count, M. Ghobakhloo is the most influential, followed by D. Ivanov and M. Javaid. The study reveals limited collaboration among authors, although notable international partnerships exist among countries such as China, India, the United Kingdom, the United States, and France. Key organizational collaborations include the Federal University of Rio Grande do Sul Organizational Engineering Group, Jamia Millia Islamia Department of Mechanical Engineering, Govind Ballabh Pant University Department of Industrial & Production Engineering, the Berlin School of Economics and Law Supply Chain and Operations Management, and Kaunas University of Technology School of Economics and Business.

Furthermore, a co-occurrence analysis identifies the most frequently associated keywords. "Industry 4.0" emerges as the most prevalent keyword, followed by "blockchain" and "digital twin."

Ultimately, this study underscores the pivotal role digital twin technology will play in shaping the future of supply chain management. By exploring the concept of digital twins, which is rapidly gaining traction across various industries, along with the enabling technologies that support its potential, this research highlights the evolving practices in supply chain management. This study fills a critical gap in the literature and provides a bibliometric foundation for guiding future research directions in digital twin technology and its application in supply chain management.

References

Casino, F., Dasaklis, T. K., & Patsakis, C. (2019). A systematic literature review of blockchain-based applications: Current status, classification and open issues. *Telematics and Informatics, 36*, 55–81. https://doi.org/10.1016/j.tele.2018.11.006

Dalenogare, L. S., Benitez, G. B., Ayala, N. F., & Frank, A. G. (2018). The expected contribution of Industry 4.0 technologies for industrial performance. *International Journal of Production Economics, 204*, 383–394. https://doi.org/10.1016/j.ijpe.2018.08.019

Dwivedi, Y. K., Hughes, L., Baabdullah, A. M., Ribeiro-Navarrete, S., Giannakis, M., Al-Debei, M. M., et al. (2022). Metaverse beyond the hype: Multidisciplinary perspectives on emerging challenges, opportunities, and agenda for research, practice and policy. *International Journal of Information Management, 66*, 102542. https://doi.org/10.1016/j.ijinfomgt.2022.102542

Frank, A. G., Dalenogare, L. S., & Ayala, N. F. (2019). Industry 4.0 technologies: Implementation patterns in manufacturing companies. *International Journal of Production Economics, 210*, 15–26. https://doi.org/10.1016/j.ijpe.2019.01.004

Ghobakhloo, M. (2018). The future of manufacturing industry: A strategic roadmap toward Industry 4.0. *Journal of Manufacturing Technology Management, 29*(6), 910–936. https://doi.org/10.1108/JMTM-02-2018-0057

Ghobakhloo, M. (2020). Industry 4.0, digitization, and opportunities for sustainability. *Journal of Cleaner Production, 252*, 119869. https://doi.org/10.1016/j.jclepro.2019.119869

Ivanov, D. (2020). Predicting the impacts of epidemic outbreaks on global supply chains: A simulation-based analysis on the coronavirus outbreak (COVID-19/SARS-CoV-2) case.

Transportation Research Part E: Logistics and Transportation Review, 136, 101922. https://doi.org/10.1016/j.tre.2020.101922

Ivanov, D., Dolgui, A., & Sokolov, B. (2021). Cloud supply chain: Integrating and configuring digital twins in supply chain risk analytics. *International Journal of Production Research, 59*(1), 45–61. https://doi.org/10.1080/00207543.2020.1824082

Kamble, S. S., Gunasekaran, A., & Dhone, N. C. (2018). Sustainable Industry 4.0 framework: A systematic literature review identifying the current trends and future perspectives. *Process Safety and Environmental Protection, 117*, 408–425. https://doi.org/10.1016/j.psep.2018.05.009

Kurutkan, M., & Orhan, F. (2018). Sağlık Politikası Konusunun Bilim Haritalama Teknikleri ile Analizi. In *Bilim Haritalama, Bibliyometrik Analiz ve Kitap İle İlgili Genel Hususlar* (pp. 2–12). IKSAD Publishing.

Maddikunta, P. K. R., Pham, Q.-V., Prabadevi, B., Deepa, N., Dev, K., Gadekallu, T. R., et al. (2022). Industry 5.0: A survey on enabling technologies and potential applications. *Journal of Industrial Information Integration, 26*, 100257. https://doi.org/10.1016/j.jii.2021.100257

Oztemel, E., & Gursev, S. (2020). Literature review of Industry 4.0 and related technologies. *Journal of Intelligent Manufacturing, 31*(1), 127–182. https://doi.org/10.1007/s10845-018-1433-8

Tao, F., Zhang, H., Liu, A., & Nee, A. Y. C. (2019). Digital twin in industry: State-of-the-art. *IEEE Transactions on Industrial Informatics, 15*(4), 2405–2415. https://doi.org/10.1109/TII.2018.2873186

Chapter 3
Harnessing the Power of Simulation in Supply Chain Digital Twins

Emrah Koksalmis

Abstract Digital twin (DT) plays a vital role in driving digital transformation across varied sectors, including supply chains, manufacturing, logistics, etc. In supply chain management, DTs offer significant benefits by providing detailed virtual models of supply chain processes. This allows real-time monitoring, predictive analysis, and optimization, leading to enhanced efficiency, cost reduction, and improved service levels. Despite common misconceptions, DTs and simulation models are distinct but complementary, working together to provide comprehensive insights into system behavior and performance. This chapter explores key studies on the application of simulation and DT in supply chain management, clarifying concepts, examining implementation, and reviewing exemplary applications. The conclusion highlights key challenges and future directions for DT and simulation models in supply chain management.

Keywords Supply chain management · Digital twin · Simulation · Digital transformation

3.1 Introduction

Digital Twin (DT) technology has emerged as a transformative power across various businesses, offering new ways to optimize operations, predict outcomes, and enhance decision-making processes. By generating virtual replicas of physical entities and systems, DTs provide detailed insights and enable real-time monitoring and analysis. Although the origins of the DT term can be traced back to early theoretical models and their practical application by Shafto et al. (2010) in NASA's draft technological roadmap, the idea of the DT originated from a University

E. Koksalmis (✉)
Department of Industrial Engineering, Turkish Air Force Academy, National Defense University, Istanbul, Türkiye
e-mail: emrah.koksalmis@msu.edu.tr

G. Rabadi, B. Soykan (eds.), *Optimizing Supply Chains Through Digital Twins*, International Series in Operations Research & Management Science 375,
https://doi.org/10.1007/978-3-032-08147-6_3

of Michigan presentation by Grieves and Vickers (2017) aimed at establishing a Product Lifecycle Management center. Today, DTs are integral to driving digital transformation, supporting a wide array of applications from manufacturing and logistics to healthcare and urban planning. As advancements in technologies like machine learning and big data continue to evolve, the potential for DT to revolutionize industries and create significant economic value becomes increasingly apparent.

According to Global Market Insights (2020), the DT market exceeded $5 billion in 2020 and is projected to increase at an annual growth rate of over 35% from 2021 to 2027. In their report for McKinsey & Company, Chui et al. (2021) estimate that by 2030, DT technology could generate global economic value between $5.5 trillion and $12.6 trillion. Although these estimates vary significantly, it is obvious that DT will be crucial in driving digital transformation initiatives for businesses in the coming years, as stated by Biller et al. (2023).

DT offers substantial benefits for supply chain management by providing a detailed virtual model of physical supply chain processes. This allows for real-time monitoring, predictive analysis, and the optimization of logistics operations. By aggregating data from multiple sources, DTs can improve visibility across the supply chain, foresee potential disruptions before they happen, and support informed decision-making. This results in enhanced efficiency, cost reduction, and improved service levels. Additionally, DTs can simulate various scenarios, enabling businesses to assess the effects of different strategies and make proactive adjustments, thus enhancing resilience and adaptability in a complex and dynamic market environment.

In their survey about simulation and supply chain, Terzi and Cavalieri (2004) state that despite the extensive use of IT solutions in supply chain management, the industry still faces significant challenges due to the inherent complexity of logistics network problems and the conflicts between localized goals and broader network strategies. In a multi-decisional supply chain context, simulation is a vital technique, particularly for conducting what–if analyses. These analyses provide a quantitative evaluation of the benefits and issues of operating cooperatively rather than through solely transactional relationships with upstream and downstream tiers. Consequently, simulation knowledge is considered one of the key competencies for modern enterprises across various fields, including supply chains, marketing, manufacturing, etc. Many organizations regard simulation as a critical decision support system.

According to Wright and Davidson (2020), simulation is frequently combined with DT technology, leading to a common misconception that simulation models and DTs are the same. Many organizations equate DTs with simulation models, and numerous software and consulting companies produce and market simulation software under the guise of being DTs. DT and simulation are an inseparable pair, working together to provide comprehensive and dynamic insights into system behavior and performance. While both simulation and DT models replicate a physical system in a virtual environment, they are distinct from one another. However, simulation and DT are so closely linked in discussions and applications that they are often seen as inseparable, as stated by Wooley et al. (2023). This observation is supported by various studies in the literature. Many researchers do not clearly distinguish between DTs and simulation models in their work. In fact, a significant number of papers develop simulations and label them as DTs, although it is unclear whether these simulations have the full capabilities typically associated with DTs.

This chapter delves into pivotal studies on the application of simulation and DT in supply chain management. The remaining sections of this chapter are organized as follows: Sect. 3.2 clarifies the concepts of DT and simulation, addressing the often-ambiguous distinctions between these two. Section 3.3 examines the development and implementation of supply chain DTs, emphasizing essential components and critical questions necessary for effective creation, with a focus on the central role of simulation. Section 3.4 reviews exemplary studies that accurately apply DTs in supply chain management, showcasing the effectiveness of real-time data and integrated frameworks for simulation and optimization. Section 3.5 offers a conclusion, highlighting key challenges and future directions in the development and application of simulation models within supply chain DTs.

3.2 Digital Twin and Simulation

The idea of the DT originated from a 2002 University of Michigan presentation by Grieves and Vickers (2017) aimed at establishing a Product Lifecycle Management center. The slide, titled "Conceptual Ideal for PLM," involved key components of the DT: real space, virtual space, data flow links from real to virtual space, information flow links from virtual to real space, and virtual sub-spaces.

However, the term "Digital Twin" first emerged in Shafto et al. (2010) in NASA's draft technological roadmap, where it was also referred to as "Virtual Digital Fleet Leader." NASA pioneered the definition of DT, describing it as "an integrated multi-physics, multi-scale, probabilistic simulation of a vehicle or system that uses the best available physical models, sensor updates, fleet history, etc., to mirror the life of its flying twin." Although the term was first mentioned in NASA's 2010 roadmap, the agency had previously applied a similar concept during the Apollo program in 1960s as stated by Rosen et al. (2015). In the Apollo program, two identical space vehicles were constructed to mirror each other. Following NASA's example, the US Air Force quickly applied DT technology to the design, maintenance, and prediction of their aircraft and published a conceptual model in a study by Tuegel et al. (2011).

To establish a consensus on the definition of a DT, it is vital to outline its fundamental requirements and capabilities. According to Wooley et al. (2023), these specifications have evolved over time, driven by technological advancements in areas such as machine learning and big data. Despite the ongoing lack of consensus on the definition of the term DT, the most widely accepted definition of a DT has been provided by The Digital Twin Consortium (2020). They describe DT as the virtual representations of real-world entities and processes synchronized at a specified frequency and fidelity. For instance, a real-world entity might be an industrial machine, whose virtual counterpart is known as an "asset twin," replicating the behavior of the physical asset. According to Biller et al. (2023), DT can also model entire systems of physical assets, such as a factory or a supply chain network that includes suppliers, factories, customers, etc.

DT technology, with its origins tracing back to early 2000s concepts and NASA's pioneering efforts, has established itself as a pivotal element in modern industry and technology. The extensive projections for its economic impact highlight the increasing relevance and potential of DTs in enhancing efficiency and innovation. By creating virtual representations of real-world entities and systems, DTs offer incomparable opportunities for all businesses. As technological advancements in machine learning and big data continue to drive the evolution of DT capabilities, it is essential to outline and standardize their fundamental requirements to fully harness their potential. The continued integration and adoption of DTs will undoubtedly play a significant role in shaping the future of various sectors, including supply chain, contributing to sustained growth and technological advancement.

DT presents significant opportunities for supply chain management by providing a comprehensive virtual representation of physical supply chain processes. This enables real-time monitoring, predictive analysis, and optimization of logistics operations. By integrating data from various sources, DTs can enhance visibility across the supply chain, identify potential disruptions before they occur, and facilitate informed decision-making. This leads to improved efficiency, reduced costs, and higher service levels. Moreover, DTs can simulate different scenarios, allowing businesses to evaluate the impact of various strategies and adjust proactively, thus fostering resilience and adaptability in an increasingly complex and dynamic market environment.

Despite the widespread adoption of various IT solutions in supply chain management, the industry continues to confront substantial challenges. These challenges are primarily attributed to the natural complexity of logistics network problems and the discord between localized objectives and broader network strategies. In this multifaceted and decision-intensive supply chain context, simulation has emerged as a pivotal technique. It is particularly valued for its capacity to perform what–if analyses, which enable the rigorous, quantitative assessment of the potential benefits and issues associated with operating within a cooperative framework rather than solely relying on transactional relationships with upstream and downstream tiers. According to Terzi and Cavalieri (2004), these simulations provide crucial insights that facilitate more informed and strategic decision-making in supply chain management.

In the industrial sector, simulation has long served as a vital tool for production engineers. Simulation knowledge is highly regarded as one of the essential competencies for modern enterprises across various processes, including supply chain, marketing, manufacturing, etc. Many organizations view simulation as an indispensable decision support system. As stated by Terzi and Cavalieri (2004), since 1996, the USA Department of Defense (DoD) has required all its services and parts suppliers to provide a simulation model of the products or services they offer.

DTs and simulation are an inseparable duo, working together to provide comprehensive and dynamic insights into system behavior and performance. Therefore, simulation has been frequently highlighted as a crucial concept, owing to its vital role in developing DTs and aiding in management decision-making. According to Mesquita et al. (2024), this phenomenon has also been observed in various studies within the literature. In their study, where words and expressions were examined using Python's wordcloud package, simulation emerged as the most frequently cited concept, highlighting its essential role in the development of DT and supporting management decision-making.

Wooley et al. (2023) state that simulation and DT are concepts used together to such an extent that they are almost inseparable in discussions and applications. This phenomenon has also been observed in various studies within the literature. Many researchers have not clearly differentiated between DT and simulation models in their studies. In fact, a significant number of papers build simulations and refer to them as DTs, but it remains ambiguous whether these simulations possess the full capabilities typically associated with DTs.

Wright and Davidson (2020) emphasize that, while simulation and DT models both replicate a physical system in a virtual environment, they are distinct from each other. Ante (2021) states that simulation is often combined with DT technology, leading to a widespread misconception that simulation models are the same as DTs and vice versa. Furthermore, Wooley et al. (2023) claim that many organizations equate the term DT with simulation models, and numerous software and consulting companies are producing and marketing simulation software under the guise of being DT.

Simulation models are frequently employed for a general understanding of systems and for making generalized predictions. However, they rarely represent a system's real-time status accurately due to the absence of real-time data. This limitation renders these models or simulations static, as they do not update or generate new predictions without being fed by new information. Besides, as stated by Singh et al. (2021), real-time data alone is insufficient for a simulation to function as a DT effectively. Data must be automatically loaded into the DT, and there should be a bidirectional flow between the physical and digital realms. Although some studies in the literature claim to use real-time data connections in their case studies, they rely on historical data, highlighting the challenges of implementing real-time data transfer. Additionally, according to Wooley et al. (2023), many DT-related papers are often vague about data implementation details and do not provide specific guidance that other researchers can use.

In their study, Mesquita et al. (2024) show that simulated data is one of the data sources used in DT studies in the literature. Figure 3.1 illustrates the various data sources used in the DT case studies analyzed in this study. Most of the data in these studies are secondary, with primary sources being datasets that store historical sensor data, inventory levels, customer transactions, and customer location/navigation data, and real-time data are sourced from sensors, cameras,

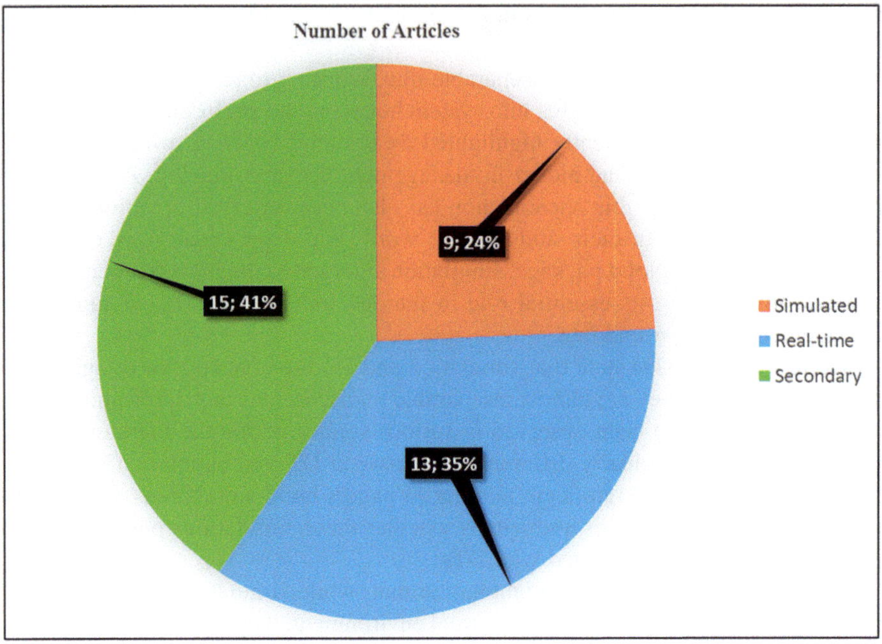

Fig. 3.1 Types of data used in the case studies (Mesquita et al., 2024)

databases, and platforms. Despite not aligning with the definition of a DT in terms of data use and transfer, some case studies, which are identified as DT by the authors themselves, use simulated data created by the authors using tools like MATLAB, Excel, and R Programming Language.

In the literature, the relationship between simulation and DT has been formulated in various ways. For instance, Ivanov and Dolgui (2019) describe a DT as a triad of simulation, optimization, and data analytics. They also highlight key technologies for supply chain DT, including RFID, the internet of things, track and trace systems, and online risk databases. According to Barykin et al. (2020), this emphasizes the multifaceted nature of DTs and their integration with simulation, optimization, and data analytics. It underscores the critical technologies that enable the functioning of supply chain DTs, highlighting the comprehensive and interconnected approach required to effectively implement and utilize DTs in supply chain management.

With the advent of DTs, the role of simulation has expanded further, providing dynamic and comprehensive virtual representations of physical systems that integrate real-time data. This integration enhances the efficiency and resilience of supply chains, fostering innovation and adaptability in a complex market landscape. The close relationship between simulation and DTs, often conflated yet distinct, underscores their combined transformative potential in optimizing supply chain operations. The literature reveals that simulation and DT concepts are used together to such an extent that they are almost inseparable in discussions and applications.

This phenomenon has been widely observed and acknowledged in various studies, highlighting the crucial role of simulation in the development and implementation of DTs. The continued integration and adoption of DTs will undoubtedly play a significant role in shaping the future of supply chain, contributing to sustained growth and technological advancement. Therefore, the next section will closely examine the relationship between Supply Chain DT and simulation.

3.3 Simulation in Supply Chain Digital Twin

The development of supply chain DTs aims to address and mitigate risks within supply chains, thereby improving their reliability and sustainability amid potential disruptions. Barykin et al. (2020) state that this supports the evolution of modern supply chains from traditional, linear models to more interconnected and synchronized networks of trading partners. This shift toward a cohesive and integrated approach has been foundationally described in studies such as those by De Souza et al. (2017), highlighting the transformation toward enhanced collaboration and real-time responsiveness among supply chain participants.

To effectively address the challenges and opportunities mentioned in the previous paragraph, the employment of sophisticated methodologies such as simulation is essential. These advanced techniques are indispensable for managing the complex and dynamic nature of contemporary supply chains. Simulation facilitates the testing and validation of various scenarios, enabling supply chain managers to anticipate potential disruptions and formulate robust mitigation strategies. By providing a detailed and dynamic virtual representation of supply chain processes, simulation enhances decision-making, optimizes real-time responsiveness, and promotes improved collaboration among trading partners. These capabilities are crucial for ensuring the reliability and sustainability of supply chains in an increasingly interconnected and volatile market environment.

Although there are some other simulation companies working in this area, the most well-known one is AnyLogic. They have been developing a comprehensive conceptual model for a supply chain DT, which integrates various methods, including scenario planning, analytic optimization techniques (e.g., Genetic Algorithm, Non-Linear Optimization Methods, etc.), simulation, and decision support tools. Barykin et al. (2020) state that these elements work together to enhance supply chain management, addressing both strategic and operational challenges effectively. Similarly, Ivanov and Dolgui (2019) conceptualize a DT through a circuit diagram that integrates Simulation, Optimization, and Data Analytics into a cohesive triad. They further define essential technologies that facilitate the implementation of supply chain DTs, including RFID, the Internet of Things, track and trace systems, and online risk databases.

Kauf (2019) suggests that it is essential to implement a broad spectrum of tools that fully automate logistics tasks, in conjunction with cutting-edge intelligent technologies and smart solutions. Key elements in this endeavor include Intelligent Transportation Systems (ITS), autonomous logistics, the physical internet, intelligent cargo, and self-organizing logistics. These advancements have markedly enhanced the tracking of commodity movement, the monitoring of warehouse and store inventories, the collection of data on item orders, delivery reliability, and overall customer satisfaction. These technologies furnish the real-time data imperative for simulation models. By providing precise and up-to-date information, they significantly enhance the capability to develop accurate virtual models of physical processes. The integration of real-time data is critical for simulations to faithfully represent current conditions and effectively forecast future scenarios. This, in turn, supports more informed decision-making and optimizes the overall performance of supply chain operations.

Based on the research conducted by Biller et al. (2022), a supply chain DT incorporates all the DT reference components provided below, including supply chain network simulation as a virtual representation component of the DT.

- *Virtual Representation:* The supply chain network simulation acts as the virtual representation.
- *Physical Representation:* The supply chain network serves as the physical representation.
- *Synchronization:* Hot starting the simulation means regularly updating model parameters and input risk profiles with the latest data to synchronize the physical and digital supply chain networks. The update frequency should align with decision-making timelines and learning needs.
- *Learning and Feedback Loop:* Insights obtained from the virtual world inform real-world supply chain management. The impact of management decisions—evidenced by actual outcomes, disruptions, IoT data, and evolving assumptions and expert opinions—is fed back into the DT as new input data or adjustments to business flows. This establishes a closed feedback loop between the physical supply chain and its DT.

Keeping these components in mind, Barykin et al. (2020) suggest that the concept for a supply chain DT should address the key questions provided below. We strongly believe that accurately addressing and implementing these questions is crucial for the proper simulation of the supply chain and the effective utilization of the DT in accordance with our needs.

- What combination of simulation modeling, optimization, and data analytics constitutes the full range of technologies required to create a DT model of the supply chain?
- How accurate or detailed should the simulation model be to qualify as a supply chain DT, and which issues should it address during the modeling process?
- Which concept of the supply chain DT is the most efficient, and what is the optimal combination of analytical optimization and dynamic simulation? Two main approaches are considered: analytical optimization as a complement to dynamic simulation, and dynamic simulation as a complement to analytical optimization.

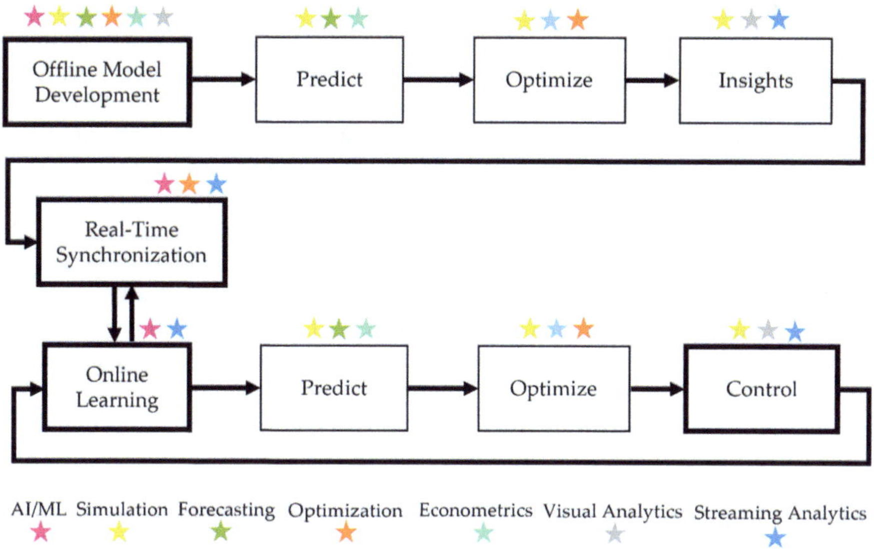

Fig. 3.2 Key factory DT functions and enabling technologies (Biller and Biller 2023)

The study by Biller et al. (2023) elucidates that the primary objectives of the DT developed for supply chains are threefold: (1) to recreate historical scenarios; (2) to gain insights into future supply chain operations by forecasting costs and service levels; and (3) to devise a strategic decision-making framework to sustain optimal supply chain performance amidst disruptions. Simulation, being a pragmatic technology for modeling, comprehending, and optimizing complex systems, constitutes a fundamental component of DTs. Leveraging the capabilities of simulation, DTs facilitate decision-making processes, thereby unlocking substantial value for businesses.

Figure 3.2 illustrates the essential functions of a factory DT, employing AI/ML, simulation, forecasting, optimization, econometrics, visual analytics, and streaming analytics as enabling technologies. As demonstrated in the figure as "yellow star," simulation is an integral technology embedded in nearly all functions of a DT developed for supply chain applications. Its comprehensive involvement underscores its critical role in modeling, understanding, and optimizing complex supply chain systems, thereby enhancing decision-making and operational efficiency.

Simulation and artificial intelligence/machine learning together are synergistic tools in the creation of DTs for complex systems. Although building stochastic factory simulations and supply chains requires significant time and effort, their critical importance in DT development is underscored by two primary capabilities: (1) uncertainty modeling and (2) explainable analytics. These capabilities highlight the essential role of factory simulations and elucidate the specific contexts in which they are indispensable.

While there are three major paradigms in simulation modeling, such as system dynamics, discrete event, and agent-based modeling, the efficiency and flexibility of discrete event simulation make it ideally suited for a wide range of supply chain applications, as highlighted by Correa Espinal et al. (2012) and Tako and Robinson (2012). For instance, within warehouse applications, discrete event simulation can be effectively utilized for evaluating performance indicators and optimizing decision-making processes. This encompasses areas such as order picking systems, product location strategies, inventory management, and warehouse design. As highlighted by Aretoulaki et al. (2024), the application of simulation is particularly suited for these complex tasks, facilitating enhanced operational efficiency and strategic planning. However, in the broader context, there is a critical need for more academic research in the fields of simulation and DT in supply chain applications since the volume of publications on this subject is remarkably low.

Although there is a growing trend toward enhancing simulation functionalities to offer real-time simulation scheduling, most of the simulation software remains process-based scheduling applications. Many existing software solutions still lack real-time capabilities, which are crucial for DT applications (Agalianos et al., 2020). This topic is beyond the scope of this chapter and will be addressed in the following chapters of this book.

The development and implementation of supply chain DT are crucial for addressing the complexities and uncertainties of modern supply chains. By integrating advanced technologies such as simulation, DT can provide a robust framework for enhanced decision-making, operational efficiency, and resilience. This evolution supports a shift toward more interconnected and synchronized supply chain networks, promoting sustainability and reliability. In this section, we provided some information about the development and implementation of supply chain DT, outline key components and questions essential for creating effective DTs, and underscore the critical role of simulation. The next section will delve into the existing studies in supply chain DT.

3.4 Related Studies in the Literature

The growing popularity of the DT concept in various industries has unfortunately led to its overuse and misapplication in academic publications. Researchers, driven by the desire for higher citation rates and greater visibility, have begun to label simulation-based studies as DT without them possessing the inherent characteristics that define true DT technology. This trend dilutes the essence of the DT concept, which is fundamentally characterized by a real-time, dynamic, and bidirectional link between a physical entity and its virtual counterpart. While each passing year brings a deeper understanding of what truly constitutes a DT, there is yet to be a universally agreed-upon definition of a DT, which adds to the confusion. The misrepresentation not only misleads readers but also hampers the integrity of scholarly discourse. It is important for the academic community to uphold rigorous standards and ensure that the term DT is reserved for works that genuinely adhere to its defining principles, thereby maintaining the credibility and scientific value of the research field.

This issue is raised by other researchers, as well. For instance, Wooley et al. (2023) systematically examine numerous publications to identify instances where the DT label has been inappropriately applied. This review reveals that many studies claiming to utilize DT lack the essential 4R capabilities (Representation—R1, Replication—R2, Reality—R3, Relational—R4) developed by Hyre et al. (2022), which include real-time data collection and bidirectional connections, classifying them as R0. About one-third of the papers reviewed fall into this category, using historical data rather than real-time inputs. The most common capability level achieved in the literature is R2 (45%), with only 8% reaching R1. Only 14% of papers achieve R3, indicating very sophisticated simulations, and none reach R4.

This discrepancy highlights a gap between theory and practice, often due to vague implementation details and the mislabeling of simulation models as DTs. Some studies use real-time data as historical data, while others create simulation–optimization platforms without direct virtual connections. The lack of synchronized real-time data in many studies limits their classification as true DTs, emphasizing the need for more rigorous standards and accurate definitions in this research field.

Given the tendency to mislabel traditional simulations as DTs, it is essential to meticulously review and analyze scholarly articles. In this section, we will present some publications from the literature that we consider good examples for simulation and DT applications in supply chain management.

One of the comprehensive studies, Lee and Lee (2021), aims to develop a DT framework for real-time logistics simulation to predict potential logistics risks and module arrival times accurately. The DT updates its virtual asset using Building Information Modeling (BIM) and Internet of Things sensors in near real-time. This virtual asset is then utilized for logistics simulation within a GIS-based routing application. The DT represents various modular construction assets such as modules, delivery trucks, workers, and other transportation objects in real time. By subscribing to real-time IoT data, the DT updates its virtual assets and tests different logistics scenarios to find potential risks and optimal delivery paths based on GIS. This approach avoids information loss by selectively exchanging necessary information for logistics monitoring and simulation at the application level. The system architecture comprises a web front end and a back-end powered by the Unity engine, which runs parallel physical simulations and supports detailed 3D simulations and data analytics for module production, logistics, and onsite assembly.

In this study, researchers suggest a DT framework's system architecture presented in Fig. 3.3. The architecture shows the overall summary of the system and its data transaction process in detail.

To test this framework, a case study was conducted by researchers on a virtual modular construction project involving a 6-story apartment with 80 modules. The project utilized synchronized physical and virtual assets, supported by hypothetical IoT sensor data, to predict logistics risks and find alternative routes in real time. The DT framework demonstrated its ability to predict risks and calculate accurate Estimated Times of Arrival (ETAs), leading to significant reductions in idle time

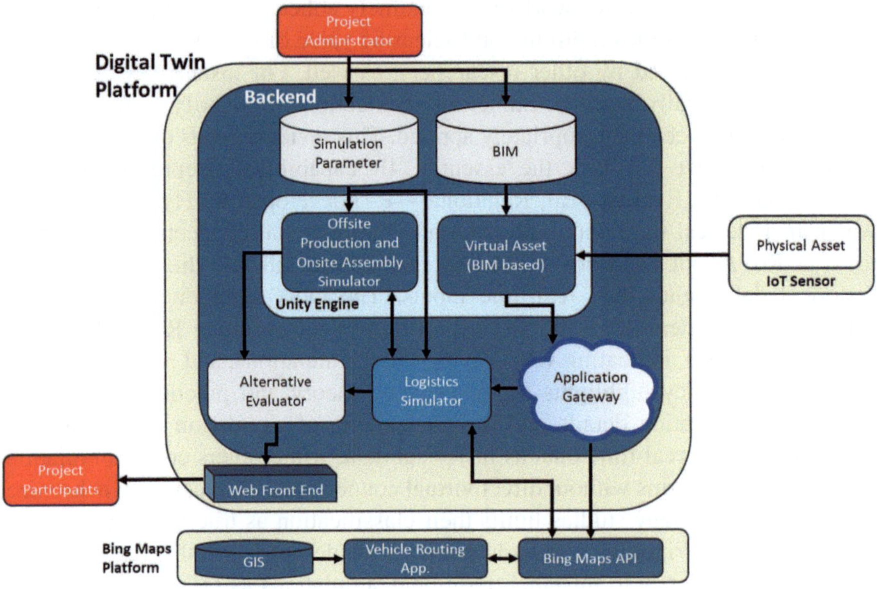

Fig. 3.3 System architecture of the DT platforms for logistic simulation—adapted from Lee and Lee (2021)

losses. The logistics simulator, connected to Bing Maps, optimized routes by considering various factors, such as avoiding permissions or tolls. The results indicate that this integrated approach can effectively coordinate and collaborate among project participants, enabling just-in-time delivery by adjusting production and delivery schedules based on accurate ETAs.

Another study, which stands out as one of the few that utilizes real-time data, is presented by Perez et al. (2022). This study presents an integrated DT framework for supply chain business processes, updated in real-time with live process data. The framework supports both offline and online optimization models: the offline mode uses deterministic optimization with recent system parameters, while the online mode creates a stochastic simulation environment for real-time optimization in a feedback loop. The framework is applied to the order-to-cash (OTC) process, which tracks customer orders from creation to payment, and incorporates digital supply chain concepts, discrete event simulation, business process management notation, queuing systems, and online optimization.

The framework consists of three main modules: System Identification, Simulation Engine, and Optimization Engine. The Simulation Engine uses SimJulia, an open-source Julia language package, to generate discrete event simulation models capturing network flow logic. Agents in the simulation represent business process servers with independent queues, processing orders through a network of tasks until fulfillment or loss. Optimization can be embedded as events within the simulation, periodically triggered or in response to system conditions. This integration allows the optimization engine to update system queues in real-time, creating a dynamic and realistic simulation environment.

In this study, the DT framework's simulation and optimization engines are illustrated with five examples, demonstrating benefits such as prioritizing customer orders, increasing revenue and on-time fulfilment, evaluating mitigation approaches, and quoting accurate lead times. The framework uses the discrete-time state-task network scheduling paradigm from the chemical process industry, integrating optimization within the simulation engine for a realistic model that accounts for both build and solution times of mixed-integer linear programming models. The study's simulation results show that the performance of heuristics is problem-specific, emphasizing the DT's role in selecting and refining optimal scheduling policies. Overall, the study highlights the potential of integrating simulation with optimization within a DT framework to enhance supply chain performance. The DT provides a powerful tool for improving supply chain operations in the digital era by offering real-time insights and adaptive optimization capabilities. This integration ensures efficient resource allocation, improved order fulfilment, and effective response to disruptions, making it a valuable asset for supply chain management.

A recent study by Abouzid and Saidi (2023) significantly contributes to the literature, demonstrating the profound impact of DT technology on improving business process efficiency and profitability. By simulating the current ("As-Is") and future ("To-Be") processes, the study evaluates key performance indicators (KPIs) like lead time, inventory levels, and delivery time. Through simulation, they observe a significant decrease in lead time of over 30%, reducing the time required to treat an order from 6 days to just 2 working days. In this study, discrete-event simulation and queue priority heuristics from queuing theory are applied to perform a Monte Carlo simulation of supply chain business processes. ProModel is used as a process simulation tool to analyze and simulate the business process. ProModel integrates with Microsoft Office Visio, allowing for the creation and execution of simulation models directly within Visio. This enabled seamless simulation of Visio flowcharts, Value Stream Maps, and workflow diagrams, which helped in predicting and enhancing the efficiency of production business processes.

In another study, Burgos and Ivanov (2021) examine the COVID-19 pandemic's impact on food retail supply chains and their resilience, using a discrete-event simulation model based on real-life scenarios in Germany. With AnyLogistix digital supply chain twin, the study focused on ten product categories across 28 supermarket locations in five countries (Germany, Austria, the Czech Republic, Italy, and Hungary) and three distribution centers in Germany. The simulation evaluated various scenarios, including increased demand, supplier shutdowns, and transport bottlenecks, to understand the overall performance and resilience of the supply chains during the pandemic. The model was validated through multiple methods, including comparing simulation results with 2020 German retail data and conducting variation experiments. The results revealed how the pandemic affected supply chain operations and performance, showing both challenges, such as increased costs and delays, and opportunities due to rising demand.

The study's findings emphasize the importance of supply chains resilience and provide a framework for post-disruption recovery. This framework includes recommendations for digitalization, inventory management, supply chain flexibility, supply chain collaboration, and end-to-end visibility. The simulations highlighted a positive correlation between the duration of disruptions and their impact on supply chains, offering insights into the synergistic effects of pandemic-related events. However, it is noted that the use of secondary data sources in the simulations may lead to potential inaccuracies. These insights aim to help food retailers improve their supply chain strategies to better withstand future disruptions and maintain competitiveness.

Ivanov and Dolgui (2021) introduce a digital supply chain twin, a real-time computerized model that integrates simulation, optimization, and data analytics to proactively and reactively manage disruption risks by building scenarios and identifying disruptions using historical and real-time data in their study. This study introduces a digital supply chain twin—a computerized model representing network states in real-time. The model integrates simulation, optimization, and data analytics to manage disruption risks. In the proactive stage, data analytics builds realistic disruption scenarios using historical disruption data and supplier reliability data from ERP systems during the supply chain design phase. At the reactive stage, data analytics identifies disruptions in real-time using process feedback data from sensors, T&T, and RFID, embedding this data into a reactive simulation model for recovery policy simulation and optimization.

The study uses anyLogistix software for developing the digital supply chain twin, combining it with data analytics to map risk data with geographical locations in the supply chain structure. During the resilient supply chain design stage, disruption risk data is used to assess supplier and transportation risks, predict supply chain interruptions, and compute alternative supply network topologies with estimated times of arrival. Real-time simulation analyzes disruption impacts and alternative supply chain designs based on current inventory, demand, and capacity data. Simulation results are transferred to ERP systems or business intelligence tools for performance impact analysis, potentially activating BI algorithms to resolve issues.

A multi-step analysis procedure involves collecting historical risk data, setting up disruption scenarios, stress-testing supply chain designs, and simulating recovery policies. For example, a natural disaster disruption like Typhoon Mangkhut in Hong Kong is simulated to observe its impact on supply chain performance. Recovery policies are simulated using real-time data updates on capacity, inventory, and lead times, with results transferred to ERP or business intelligence tools for KPI impact analysis. The system visualizes supply chain risks, assesses supplier risks, predicts supply interruptions, and computes alternative routes. While the study presents a comprehensive DT framework for supply chain resilience, it does not cover technical requirements for data processing capacities or a detailed technical analysis of disruption data filtering.

In this section, we present some publications from the literature that we consider good examples for simulation and DT applications in supply chain management. However, it is obvious that there is a noticeable scarcity of publications related to

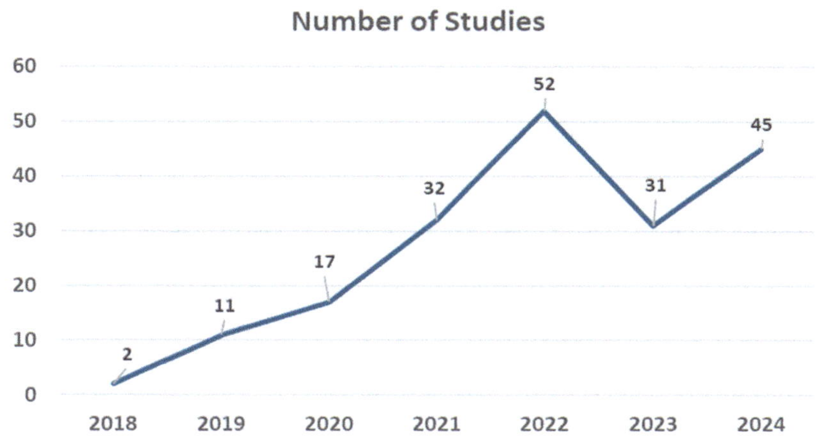

Fig. 3.4 Publication trends according to years

the topics of simulation and DT within the academic literature of supply chain, and the existing body of research remains relatively limited. While this scarcity of publications underscores the necessity for more scholarly attention, there has been a noticeable increase in the number of studies in recent years, as illustrated in Fig. 3.4.

Data collection for this figure is carried out on 2 January 2025, based on studies from the Web of Science (WoS) database, using a query of the keywords '("Simulation") AND ("Digital Twin*") AND (Supply Chain*). While "Simulation" and "Digital Twin" keywords are applied to the topic of the indexed article, "Supply Chain" is applied to all fields. The asterisk (*) at the end of a keyword guarantees the presence of the term in both singular and plural forms. Further results added thereafter to the WoS database are not included in the present study.

The initial search generates 190 results. The earliest study in this domain is published in 2018. Notably, the number of studies has shown a significant upward trend, with the maximum annual publication count reaching 52 studies in 2022. The trajectory began with 2 studies in 2018 and has risen steadily, averaging 43 studies annually over the past 3 years, which means 67% of the studies are conducted in the past 3 years.

This trend is expected to continue growing, driven by the rising interest in optimizing supply chain processes and enhancing resilience through advanced digital technologies. As industries increasingly recognize the value of DTs for real-time data integration and simulation-based decision-making, the research in this field is anticipated to expand further, providing deeper insights and innovative solutions for supply chain management.

3.5 Conclusion, Challenges, and Future Directions

In conclusion, DT technology has emerged as a transformative force across various industries, offering innovative approaches to optimize operations, predict outcomes, and enhance decision-making. By creating virtual replicas of physical entities and systems, DT facilitates detailed insights, real-time monitoring, and analysis. Originating from NASA's early theoretical models and practical applications by Shafto et al. (2010), DT now plays a crucial role in digital transformation across supply chain, manufacturing, healthcare, urban planning, etc.

The market for DT is experiencing rapid growth, with substantial economic value creation projected. Despite varying estimates, it is evident that DT will be essential in driving business digital transformation in the coming years, with advancements in machine learning and big data further enhancing DT capabilities and applications. A consistent definition of DT, as provided by Digital Twin Consortium (2020), remains vital.

Simulation and DT technology, while often used interchangeably, are distinct. Simulation offers generalized system understanding, whereas DT integrates real-time data for dynamic modeling. Literature frequently blurs these distinctions, using historical data instead of real-time inputs. Therefore, especially in supply chain management, technologies like IoT, RFID, and intelligent transportation systems are critical for accurate real-time data integration.

The development of supply chain DTs aims to enhance reliability and sustainability by addressing and mitigating risks through interconnected and synchronized networks of stakeholders. Advanced methodologies such as simulation are essential for managing complex supply chains, allowing for scenario testing, disruption anticipation, and robust mitigation strategies. Comprehensive models integrating simulation, optimization, and data analytics, supported by technologies such as RFID, IoT, and track and trace systems, are crucial for accurate virtual models and improved decision-making.

The research on DT and simulation applications in supply chain management has exhibited a remarkable upward trajectory since the earliest study published in 2018, as presented in Fig. 3.4. The comprehensive review of the Web of Science database indicates a total of 190 studies, with a significant increase in publication activity over the past 3 years, averaging 43 studies annually. This surge, accounting for 67% of the total studies, underscores the growing recognition of the pivotal role DT and simulation technologies play in enhancing supply chain processes. The peak in research activity, with 52 studies published in 2022, further highlights the escalating interest and investment in this domain.

In summary, the continued integration and adoption of DTs and simulation will significantly shape the future of supply chain management, contributing to sustained growth and technological advancement. By exploring various frameworks, practitioners can gain insights into methodologies, applications, and benefits, thereby choosing suitable frameworks or developing problem-specific solutions to enhance decision-making, operational efficiency, and resilience in supply chains.

3.5.1 Challenges and Future Directions

- The increasing popularity of the DT concept has led to its overuse and misapplication in academic publications, with many researchers labeling simulation-based studies as DTs without meeting the essential characteristics. This discrepancy between theory and practice is often due to vague implementation details and the mislabeling of simulation models. Rigorous standards, specifications, and accurate definitions are needed to maintain the integrity of the research field.
- Standardization remains a significant challenge in DT development. However, efforts to modularize the process aim to transfer learning from one project to another, reducing development time. Creating a robust simulation data model requires considerable expertise and experience, and even experts find it challenging to integrate these models into a comprehensive supply chain network simulation. As it is also stated by Biller and Biller (2023), managing the DT simulation data model in coordination with all stakeholders throughout the product and process lifecycle is a key challenge in process DT development.
- Mesquita et al. (2024) emphasize that processing vast quantities of data is inherently complex and requires significant time. Utilizing advanced statistical and machine learning methods is crucial to derive meaningful insights from this data. These techniques, however, are often intricate and challenging, demanding a deep understanding of their algorithms and the ability to handle large, diverse datasets effectively.
- The proportion of costs allocated to DT and simulation model development should be determined by the number of performance factors targeted for improvement and the complexity level desired. The needs might be fulfilled through standard IoT networks and traditional cloud database solutions. Therefore, as it is already suggested by Aretoulaki et al. (2024), it is essential for practitioners to stay open to alternative methods.
- According to Mesquita et al. (2024), many companies exhibit reluctance toward adopting digital technologies, primarily due to the high degree of heterogeneity within their systems and the existing perceptions or biases they hold. This hesitancy is often rooted in the complexity of integrating new technologies with their diverse and already-established processes. Furthermore, there's pervasive skepticism about the tangible benefits and return on investment of these technologies. These factors collectively create a significant barrier to digital transformation, making it challenging for organizations to embrace and implement digital solutions such as DT and simulation.
- Another significant challenge in the operationalization of DT and simulation is related to internet connectivity and data security as it is highlighted by Biller and Biller (2023) in the DTCPT under the category of trustworthiness. Ensuring robust internet connectivity and safeguarding data are essential to maintaining the integrity and reliability of DT systems.
- Despite the many demonstrated benefits of DTs, the impact of the transformation process is often overlooked, hindering the evaluation of their effectiveness and

areas for improvement. The lack of a standardized evaluation methodology results in inconsistent and non-comparable results. Adopting a uniform approach would offer a comprehensive understanding of DT effectiveness, providing valuable data for decision-makers about the role of simulation in DT applications. This standardization would also ensure more accurate execution and achievement of expected simulation benefits.

- Biller and Biller (2023) present another challenge for simulation and DT. Data acquired from management systems needs to be meticulously analyzed to determine some parameters, such as entity flows, setup times, maintenance intervals, and downtimes, which are crucial for building accurate simulations. Although this process is laborious and often overlooked, it is essential for ensuring the validity and success of the DT project. Properly converting system data into simulation input processes significantly enhances the accuracy and reliability of the DT.
- While Fig. 3.4 clearly illustrates a significant increase in the number of articles published in recent years, it is obvious that there is a noticeable scarcity of publications related to the topics of simulation and DT within the academic literature to conduct a systematic literature review and bibliometric analysis. Despite the growing interest and potential benefits these technologies offer in various industries, the existing body of research remains relatively limited. This scarcity of publications underscores the necessity for increased scholarly attention and investigation to fully explore and harness the capabilities of simulation and DT technologies.

Acknowledgments The foundation of this study was laid during the research conducted under the TÜBİTAK 2219—International Postdoctoral Research Fellowship Program.

References

Abouzid, I., & Saidi, R. (2023). Digital twin implementation approach in supply chain processes. *Scientific African, 21*, e01821.

Agalianos, K., Ponis, S. T., Aretoulaki, E., Plakas, G., & Efthymiou, O. (2020). Discrete event simulation and digital twins: Review and challenges for logistics. *Procedia Manufacturing, 51*, 1636–1641.

Ante, L. (2021). Digital twin technology for smart manufacturing and industry 4.0: A bibliometric analysis of the intellectual structure of the research discourse. *Manufacturing Letters, 27*, 96–102.

Aretoulaki, E., Ponis, S. T., Plakas, G., & Tzanetou, D. (2024). (DT4Smart) discrete event simulation and Digital Twins in warehouse logistics: A bibliometric and content analysis-based systematic literature review. *International Journal of Computer Integrated Manufacturing*, 1–28.

Barykin, S. Y., Bochkarev, A. A., Kalinina, O. V., & Yadykin, V. K. (2020). Concept for a supply chain digital twin. *International Journal of Mathematical, Engineering and Management Sciences, 5*(6), 1498–1515.

Biller, B., & Biller, S. (2023). Implementing digital twins that learn: AI and simulation are at the core. *Machines, 11*(4), 425.

Biller, B., Jiang, X., Yi, J., Venditti, P., & Biller, S. (2022). Simulation: The critical technology in digital twin development. In *2022 Winter simulation conference (WSC)* (pp. 1340–1355). IEEE.

Biller, S., Venditti, P., Yi, J., Jiang, X., & Biller, B. (2023). Simulation-driven digital twins: The DNA of resilient supply chains. In *2023 Winter simulation conference (WSC)* (pp. 1463–1477). IEEE.

Burgos, D., & Ivanov, D. (2021). Food retail supply chain resilience and the COVID-19 pandemic: A digital twin-based impact analysis and improvement directions. *Transportation Research Part E: Logistics and Transportation Review, 152*, 102412.

Chui, M., Collins, M., & Patel, M. (2021). The Internet of things: Catching up to an accelerating opportunity.

Correa Espinal, A. A., G6mez Montoya, R. A., & Sánchez Alzate, J. A. (2012). Improvement of operations of picking and dispatch for a business in the mattress industry, supported by discrete simulation. *Dyna, 79*(173), 104–112.

De Souza, R., William, L., & Timperio, G. (2017). *Supply chain digital transformation: Insights and tools–anylogistix supply chain optimization software.*

Global Market Insights. (2020). Digital twin market size by application, by end use, industry analysis report, regional outlook, growth potential, competitive market share & forecast, 2021–2027. *Report ID GMI2196.*

Grieves, M., & Vickers, J. (2017). Digital twin: Mitigating unpredictable, undesirable emergent behavior in complex systems. *Transdisciplinary perspectives on complex systems: New findings and approaches*, 85–113.

Hyre, A., Harris, G., Osho, J., Pantelidakis, M., Mykoniatis, K., & Liu, J. (2022). Digital twins: Representation, replication, reality, and relational (4Rs). *Manufacturing Letters, 31*, 20–23.

Ivanov, D., & Dolgui, A. (2019). New disruption risk management perspectives in supply chains: Digital twins, the ripple effect, and resileanness. *IFAC-PapersOnLine, 52*(13), 337–342.

Ivanov, D., & Dolgui, A. (2021). A digital supply chain twin for managing the disruption risks and resilience in the era of Industry 4.0. *Production Planning and Control, 32*(9), 775–788.

Kauf, S. (2019). Smart logistics as a basis for the development of the smart city. *Transportation Research Procedia, 39*, 143–149.

Lee, D., & Lee, S. (2021). Digital twin for supply chain coordination in modular construction. *Applied Sciences, 11*(13), 5909.

Mesquita, R. P., Leal, F., & De Queiroz, J. A. (2024). Digital twins in the retail industry: A systematic literature review. *International Journal of Simulation Modelling (IJSIMM), 23*(3).

Perez, H. D., Wassick, J. M., & Grossmann, I. E. (2022). A digital twin framework for online optimization of supply chain business processes. *Computers and Chemical Engineering, 166*, 107972.

Rosen, R., Von Wichert, G., Lo, G., & Bettenhausen, K. D. (2015). About the importance of autonomy and digital twins for the future of manufacturing. *Ifac-papersonline, 48*(3), 567–572.

Shafto, M., Conroy, M., Doyle, R., Glaessgen, E., Kemp, C., LeMoigne, J., & Wang, L. (2010). Draft modeling, simulation, information technology & processing roadmap. *Technology Area, 11*, 1–32.

Singh, M., Fuenmayor, E., Hinchy, E. P., Qiao, Y., Murray, N., & Devine, D. (2021). Digital twin: Origin to future. *Applied System Innovation, 4*(2), 36.

Tako, A. A., & Robinson, S. (2012). The application of discrete event simulation and system dynamics in the logistics and supply chain context. *Decision support systems, 52*(4), 802–815.

Terzi, S., & Cavalieri, S. (2004). Simulation in the supply chain context: A survey. *Computers in Industry, 53*(1), 3–16.

Tuegel, E. J., Ingraffea, A. R., Eason, T. G., & Spottswood, S. M. (2011). Reengineering aircraft structural life prediction using a digital twin. *International Journal of Aerospace Engineering, 2011*(1), 154798.

Wooley, A., Silva, D. F., & Bitencourt, J. (2023). When is a simulation a digital twin? A systematic literature review. *Manufacturing Letters, 35*, 940–951.

Wright, L., & Davidson, S. (2020). How to tell the difference between a model and a digital twin. *Advanced Modeling and Simulation in Engineering Sciences, 7*, 1–13.

Chapter 4
Supply Chain Digital Twin Application: Control Tower

Sean Jefferson and Rami Musa

Abstract The ability to visually track and manage supply chain operations in real time is a long-awaited dream for supply chain practitioners. By utilizing control towers, which leverage the vast amount of data available today and advancements in computational and storage technologies, we can proactively address operational issues before they escalate. This chapter explores the concept of the supply chain control tower and its core engine being a digital twin, outlining how we can develop and enhance this indispensable solution to fundamentally change the way supply chain professionals manage continuity of supply to meet demand. We delved into the various applications of control towers across supply chain functions (source, make, deliver, and plan). Four main use cases are described in detail: (1) visualizing the supply chain and it corresponding key metrics, (2) detecting metric violations through flags and events, (3) conducting what-if analysis to address multiple issues collectively, and (4) utilizing optimization techniques and machine learning for supporting with mitigation solutions to be adjusted manually or automatically reach to the level of supply chain self-healing. Additionally, we share practical best practices, such as tailoring the control tower to specific roles (personas), employing agile development practices, deploying Generative AI (GenAI), and managing change effectively. The selection of enabling technologies is also discussed, along with the challenges that may arise during implementation and how to mitigate associated risks. We conclude with insights and summary of the chapter and future trends of using digital twin in supply chain.

Important Note: Please note that the content provided in this chapter represents the opinions and viewpoints of the authors, and not those of Johnson & Johnson.

S. Jefferson (✉) · R. Musa
Johnson & Johnson, Santa Clara, CA, USA

Johnson & Johnson, West Chester, PA, USA

G. Rabadi, B. Soykan (eds.), *Optimizing Supply Chains Through Digital Twins*, International Series in Operations Research & Management Science 375,
https://doi.org/10.1007/978-3-032-08147-6_4

Keywords Supply chain · Digital twin · Control tower · Near real-time supply chain management · Predictive analytics · Scenario simulation · Optimization

4.1 Introduction

Digital Twin technology is a powerful approach for integrating real-world data into simulated environments, allowing for enhanced analysis and decision-making. By creating a virtual replica of various systems or entities, organizations can gain valuable insights into their performance and interactions. In the case of supply chain management, a Digital Twin enables companies to mirror their entire supply chain process—from sourcing raw materials to delivering finished products—within a sophisticated simulation environment. This provides a holistic view of the supply chain, facilitating the monitoring of key performance indicators (KPIs) and enabling proactive management. Here is a deeper look at how Digital Twin technology can be applied to the supply chain:

- Real-Time Monitoring: By continuously feeding real-time data into the Digital Twin, organizations can observe supply chain activities as they occur. This constant stream of information helps identify bottlenecks, delays, and inefficiencies immediately, allowing for rapid response and corrective actions.
- Scenario Simulation: The Digital Twin allows for the simulation of various scenarios, enabling supply chain managers to model different conditions and their potential impacts. For example, they can assess how changes in demand, supplier issues, or transportation disruptions would affect overall operations. This capability helps in strategizing more effectively and mitigating risks.
- Predictive Analytics: Utilizing advanced analytics and machine learning algorithms within the Digital Twin can lead to predictive insights. By analyzing historical data alongside real-time information, businesses can forecast demand fluctuations, optimize inventory levels, and anticipate disruptions, ensuring a more resilient supply chain.
- Collaborative Decision-Making: The Digital Twin acts as a shared platform for stakeholders across the supply chain. With a clear visualization of the entire operation, teams can collaborate more effectively, making informed decisions that align with overall business objectives.
- Optimization of Resources: By analyzing the simulation outcomes, organizations can identify opportunities for resource optimization. Whether it's improving logistics efficiency, reducing waste, or reallocating workforce requirements, the Digital Twin aids in aligning resources with actual demand.
- Enhanced Agility: In today's fast-paced market environment, the ability to respond quickly to changes is critical. The insights derived from a Digital Twin empower organizations to adapt their strategies swiftly, ensuring they remain agile and competitive.
- Lifecycle Management: Beyond immediate operational benefits, Digital Twin technology supports long-term lifecycle management of products and services.

Companies can analyze data over time to enhance product design, improve customer satisfaction, and refine supply chain processes.

This chapter is organized into four sections. It begins with an introduction (Sect. 4.1—this section), where we explore various types of digital models, including the digital twin. We discuss four general types of models, with a focus on the use of intelligent and hybrid twins. Additionally, we provide a foundational definition of a Control Tower as an application of digital twin technology, highlighting its benefits to the supply chain in general. In Sect. 4.2, we present a literature review on control tower applications and the use of digital twins. Given the limited references available on this topic, our insights are predominantly derived from our industry experiences. Following this, we describe the technical architecture of our solution in detail, outlining the various solution stacks from data acquisition to the provision of insights for the supply chain. Finally, we conclude the chapter with a summary and conclusions. Section 4.3 will present details and considerations for building, deploying, and use of a Supply Chain Control Tower. The chapter will conclude with a summary of the different dimensions of the use of digital twin to support a Control Tower in Sect. 4.4.

4.1.1 Types of Digital Models

To start, let us classify digital models, as shown in Fig. 4.1. Traditional digital models aim to replicate reality by relying on manually input data, processing it, and feeding the output manually as well. This category is known as the offline digital model (or offline digital model), designed for long-term analysis rather than real-time insights and adjustments. A more advanced type is the digital shadow, which automates data capture and produces manual results that require human intervention to execute recommendations. The most sophisticated model is the intelligent twin,

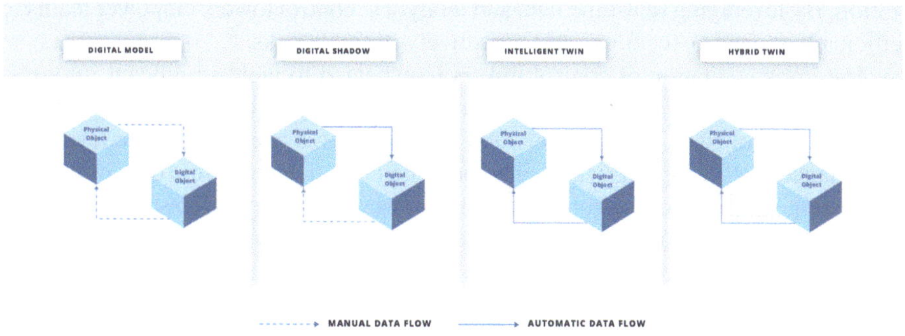

Fig. 4.1 Types of digital models differentiated by data flows. Digital models refer to modeling techniques such as simulation, machine learning (ML), and optimization

which operates in real-time and constantly exchanges data to adapt, creating closed-loop, self-regulating systems. The hybrid model, which combines elements of both the digital shadow and intelligent twin, is particularly notable due to practical reasons. This model strikes a balance between real-time capabilities and automated data generation. It proves to be a realistic and practical approach for supply chain control towers in the future.

4.1.2 What Is Control Tower?

In supply chain management, it is crucial to have proper planning for supply and demand in the short, medium, and long term. However, there are always exceptions that deviate from our planned processes related to both supply and demand for products. Handling these exceptions poses a challenge and goes beyond the typical responsibilities of a planner in a traditional supply chain.

To address this challenge, control towers have emerged as a solution at the operational level of running supply chains. Control towers help mitigate issues before they escalate, ensuring that available assets and resources are optimally managed to maintain desired service levels and optimize costs. Advanced control towers serve as command centers where a team of analysts monitors screens and are informed of potential disruptions so they can make daily changes to address issues proactively. Control Towers are a natural first step toward the creation of a digital twin of a given supply chain.

4.1.3 Control Tower Benefits

Supply chain control towers serve as centralized platforms that facilitate seamless collaboration among professionals across supply chain functions within an organization. By leveraging real-time data and analytics, control towers empower teams to efficiently manage exceptions and proactively mitigate risks.

The implementation of control towers is expected to yield significant improvements in key metrics. For instance, inventory levels can be optimized, ensuring the right amount of stock is available when and where it is needed. Service levels can be enhanced, allowing for better customer satisfaction and on-time deliveries, thus preventing stockouts or backorders. Control towers also help address the challenges associated with slow-moving and obsolete inventory (SLOB), reducing waste and improving overall operational efficiency.

To effectively achieve these improvements, control towers enable a deep understanding of the supply chain landscape. This includes monitoring and analyzing factors such as shipment delays, unexpected spikes in demand, unforeseen events like pandemics or natural disasters, disruptions in the supply chain, customs delays, geopolitical challenges, and imbalances in capacity. Having visibility into these elements allows teams to make informed decisions, take proactive measures, and dynamically adjust operations to ensure smooth supply chain operations and resilience.

In summary, supply chain control towers enable collaborative management of exceptions and risks by leveraging real-time data and insights. Their implementation drives improvements in inventory levels, service levels, SLOB, and backorders by addressing various challenges and optimizing operations in an agile manner.

4.2 Literature Review

A well-implemented Supply Chain Control Tower is multifaceted and has different capabilities to best enable the reliable supply of goods. The work by Aamir Rashid (2024) examines how information processing and the digital supply chain can help provide resilience in times of extraordinary disturbances. The authors used data collected from 251 supply chain professionals in the manufacturing industry and found that information processing capability (including visibility) and digital supply chain significantly and positively affect supply chain risk management and resilience.

When Supply Chain Control Towers are used as a central hub using technology to capture and use data to enhance visibility and improve decision-making, the authors of Abdel-Aziz Ahmad Sharabati (2022) found that Control Towers have a significant positive effect on the total competitive advantage. They also positively affect quality, cost, reliability, and responsiveness. Interestingly, they found that it does not have a significant effect on overall innovation.

Marmolejo-Saucedo (2020) found that while many players in the industry are strengthening capabilities, including major ERP (Enterprise Resource Planning) systems, there is no single technological platform capable of performing all the activities of a digital twin. They stated that digital twins consist of three components: a data model, a set of analytics or algorithms, and knowledge. In the experiment (Marmolejo-Saucedo, 2020) performed, data and advanced modeling techniques were combined to create a fit-for-purpose solution that was able to react to supply chain disruptions better than a fixed replenishment point/fixed replenishment quantity policy.

Supply chains not only rely on data and technology but must also consider the people who interact with them and utilize the information they provide to make decisions. In his paper on intelligent control towers, Vlachos (2021) performs a case study on them as a socio-technical system. The study found that interactions between technologies, the team operating the supply chain control tower, and internal and external supply chain stakeholders are critical in managing the development, implementation, and growth of intelligent supply chains. When examining the technology element, Vlachos (2021) determined that Intelligent supply chains depend on the successful integration of different technologies, including ERPs, transportation management systems, and advanced analytics executed through intelligent assets. He concluded that intelligent supply chains emerge as a result of systematically managing socio-technical interactions (Vlachos, 2021).

While each piece of literature reviewed focused on different aspects, all found evidence that Control Towers, especially those that integrating advanced modeling, have a significant positive impact on the overall health of a supply chain. However, there were limited publications addressing how to implement a control tower, including design considerations and user interaction with the system. The transcript below begins to explore perspectives on deploying a Control Tower in a large corporation and how it can evolve toward becoming a digital twin of the physical supply chain.

4.3 Solution Description

Figure 4.2 illustrates a framework for a digital twin solution addressing the supply chain control tower. There are nine layers in this stack with four foundational data layers (1–4) and five application layers (5–9). The chart depicts a hybrid intelligent twin for the supply chain that is also explained in Fig. 4.1. This is due to the input and output data flow representations. The chart illustrates the transition from the physical supply chain at the bottom to a digitalized supply chain at the top. The digitalization process involves acquiring data from various siloed IT systems, such as ERP, WMS, TMS, and others, which are primary sources of data for the supply chain digital twin in the control tower. To enable digitalization, the data is first acquired in its raw format and then transformed to align and integrate with other data sources. This is followed by connecting the siloed data using non-relational databases, specifically graph databases. These databases provide an effective representation of network data to be utilized by the application layers of the control tower. The application layers include visualization of the supply chain and personalized alerts that trigger actions to address issues by the assigned persona and enabling what–if scenario analysis. As digitalization progresses, the alerts become predictive, leading to recommended solutions for handling issues automatically or semi-automatically.

4.3.1 Digital Twin Technical Solution

4.3.2 Data Layers 1–4 (in Fig. 4.2)

Data Layers 1 and 2: Raw Data Ingestion and Transformation (Data Capture) (in Fig. 4.2)

Data is the foundational element of a control tower. Without robust and trusted data, a control tower will not be able to gain the needed adoption to deliver the

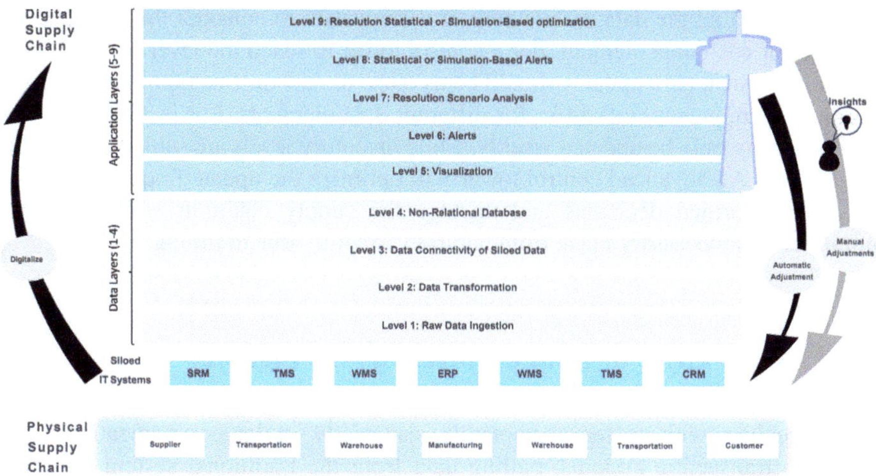

Fig. 4.2 Design framework for control tower twin stack (top five layers are application layers for the tower and the bottom four are foundational data layers)

expected value to the business. A thoughtfully created data architecture utilized a technical stack with clearly defined levels that each serve a distinct function in the aggregation, curation, and ultimate usability of data. While the latest trends are always emerging, the Medallion architecture introduced by Databricks offers one widely accepted model to curate data. Incorporating medallion data architecture or a similar alternative into the design of a control tower establishes a robust foundation for capturing data from various sources, both internal and external to the organization. This includes data from suppliers, coordination providers, ERPs (Enterprise Resource Planning), WMS (Warehouse Management System), production systems, financial systems, and other relevant sources. The architecture outlines the necessary mechanisms to collect and consolidate this data. The medallion data architecture consists of the following layers as represented in Fig. 4.2:

Bronze/raw data ingestion: unprocessed data stored as-is (e.g., direct copies from data tables in source systems).

Silver/curated and transformed data: Curated data cleansed and transformed for consistency (e.g., standardizing data fields, resolving discrepancies in naming conventions or formatting, and mapping diverse data sources into a shared schema).

Gold/aggregation: Highly refined and enriched data optimized for analytics. This layer provides a unified view of the supply chain for a specific business process by combining data from different systems, departments, and partners (i.e., supplier information, consisting of name, location, tax ID, contact name, contact phone number, contact email address, products purchased, certifications, and other relevant fields from multiple tables in potentially multiple sources).

After defining the data architecture, it is important to consider the frequency of updates for each data element. For a control tower to reach the level of capabilities of a digital twin, data updates should be as close to real-time as possible. However, the update frequency will differ for different data elements. For instance, demand forecasts may only be updated weekly, while inventory levels are subject to constant adjustment. An advanced control tower will optimize the update frequency, aiming to make informed decisions and manage the supply chain in real time while minimizing unnecessary noise from natural variability due to timing.

4.3.3 Data Layer 3: Data Connectivity (in Fig. 4.2)

When dealing with system complexity, especially in large organizations with disparate transaction systems, pulling data from the traditional system may lead to gaps. This can be the result of different systems controlling various stages of the process, as shown at the bottom of Fig. 4.2. Each system may have its data format, master data, or structure, leading to data inconsistencies and gaps when trying to integrate and consolidate information. In Fig. 4.2, this is depicted in layer 3.

In these cases, where data elements do not exactly match across systems, there are two approaches to connect the data. One is to create and maintain a fixed mapping between values in the different systems. An alternative solution is to invest in a technology that utilizes "fuzzy logic"/probabilistic matching. One example where these gaps frequently occur is in material flow. When the master data for material numbers or product names are not exact matches across systems, there is a need for stitching together the flow of a material from raw material through manufacturing, warehousing, and logistics to customers to allow end-to-end visibility and optimization.

4.3.4 Data Layer 4: Non-relational Databases (in Fig. 4.2)

For most basic visualization work, standard relational databases have been the preferred tool for storing and presenting data. With more advanced analytics, graph and vector databases may be a preferred alternative. Both systems specialize in storing highly interconnected and multidimensional data.

Both graph and vector databases provide specialized capabilities for storing and analyzing highly interconnected and multidimensional data. Graph databases excel at capturing complex relationships and their interconnections. This makes them ideal for tasks such as network analysis or recommendation systems.

On the other hand, vector databases are designed to handle high-dimensional data, such as text, images, or numerical vectors. They allow for tasks such as content-based recommendation or image recognition, in addition to supporting machine learning and simulation.

In summary, while standard relational databases have been the go-to tool for basic visualization work, more advanced analytics often require specialized databases like graph or vector databases. These technologies were developed to enable advanced machine learning and simulation. Leveraging the capabilities of these databases can enhance visualization capabilities and unlock deeper insights into complex supply chains.

4.3.5 Application Layers 5–9 (in Fig. 4.2)

Descriptive Analytics

Visualization (layer 5)—The most basic function of a control tower is making the data more easily interpreted than numbers in a system or spreadsheet through visualization. Visualization in a control tower is more than the creation of charts and graphs. It involves thoughtful consideration of which pieces of data are required to quickly identify issues or risks and act. That data then needs to be presented to the user in a format that allows for quick, unambiguous understanding, limiting the need for interpretation and facilitating quicker remediation actions. The data powering the visualizations needs to be as close to real time as possible to allow the user to have a picture of the current state of health of the supply chain. If designed well, visualization should help end users easily identify deviation from the expected performance of an end-to-end supply chain.

Diagnostic Analytics

Alerts (layer 6)—As control towers ingest and visualize huge volumes of data, the user is tasked with the interpretation of the data before they can identify risk/opportunities and finally act. Alerting helps focus users' attention to where it is most needed, allowing for management by exception. Simple alerts can be set to pre-defined thresholds (i.e., inventory at X% above or below target, demand increased more than Y% week over week). More advanced alerts can utilize multiple inputs and basic math to highlight future problems. An example of this could be supply and demand and providing alerts when the planned demand for each week is higher than the expected inventory.

Scenario modeling (layer 7)—Once an issue or potential disruption has been identified, mitigation actions need to be taken. Resolution scenario modeling allows users to model impacts across all points in a supply chain for a given product. A user can input multiple potential solutions to mitigate an issue and identify which options best achieve the desired outcome. A simple example of this would be the reallocation of inventory from one market to another. Resolution scenario planning can consider planned demand and replenishment to identify if the reallocation may have unintended consequences to other points in the supply chain.

Predictive and Prescriptive Models

Referring to Fig. 4.2, we have characterized that the advanced algorithm of the tower will be on the top two layers, i.e., predictive model alerts (layer 8) and model-based optimization (layer 9).

As opposed to the alerts and scenarios described above, this functionality relies on complex models that often rely on machine learning and other techniques such as Discrete Event Simulation (DES) and Agent-Based Modeling (ABM). Simulation techniques, such as Discrete Event Simulation (DES) or Agent-Based Modeling (ABM), along with Machine Learning (ML) methods, can be leveraged to create predictive alerts in advance using historical transactional data. DES or ABM can simulate the various events and interactions within a supply chain system, replicating real-world scenarios to forecast potential outcomes. By analyzing historical transactional data and considering factors such as demand patterns, inventory levels, and market conditions, these models can predict future alerts, highlighting potential disruptions or bottlenecks. ML algorithms, on the other hand, can learn from past data patterns to identify trends and anomalies, enabling the generation of predictive alerts. By combining the power of simulation and ML techniques, supply chain control towers can proactively identify and address potential issues, providing businesses with the ability to take preemptive actions and enhance the resilience of their supply chains.

Real-time optimization coupled with prescription and the actual adjustment of the supply chain in near real-time can significantly enhance the self-healing capabilities within the control tower. By continuously monitoring and analyzing data collected from various sources, real-time optimization algorithms can swiftly identify deviations from the desired state and prescribe immediate actions to rectify or adapt to the changing dynamics. These optimization algorithms leverage real-time and historical data to dynamically adjust inventory levels, production schedules, transportation routes, and resource allocation to optimize the performance of the supply chain. By considering factors like demand fluctuations, delivery times, and capacity constraints in real-time, they can provide recommendations for making informed decisions at the operational level. This allows supply chain managers to proactively respond to disruptions, minimize costs, maximize efficiency, and ensure continuous flow within the supply chain.

Real-time adjustment with prescription provides agility and responsiveness to unforeseen events, allowing for adaptive supply chain strategies. By constantly analyzing the data in near real-time, supply chain control towers can quickly identify issues, predict potential disruptions, and swiftly adjust their operations to mitigate risks and maintain smooth operations.

4.3.6 Stages of Maturity

We will explore two aspects of maturity: data use maturity and use cases.

(A) *Data connectivity:*

Starting with the concept of a digital shadow, we gradually transition into the realm of an intelligent or hybrid twin as depicted in Fig. 4.1. The main objective is to automate data acquisition, cleansing, and transformation, followed by establishing connections and visual representation. Through visualization techniques, we can extract valuable insights, progressing toward optimization using advanced modeling methods such as simulation and machine learning. This adjustment process can be executed either manually or automatically. When manual adjustments are made to the physical supply chain, it is referred to as the digital shadow. As we continue to gather more data and engage in scenario planning, the insights derived from this information empower us to autonomously make network changes, hence the reference to the hybrid or intelligent digital twin.

(B) *Use cases:*

Supply chain control towers mature over time through continuous improvement, thoughtful efforts to collect underlying data, and advancements in technology. In the initial stages, control towers focus on basic data visibility and alerting capabilities, providing a significant increase in near real-time visibility but delivering limited insights into the supply chain. In these beginning phases, a digital twin of the supply chain is beginning to emerge but may have pockets where data is unavailable, creating blind spots or lower fidelity of information. Typically, digital twins can be introduced and developed step-by-step, facilitating implementation and maximizing their benefits (Marmolejo-Saucedo, 2020).

1. *End-to-End visualization:* This involves having a comprehensive view of the entire supply chain in scope, allowing for better visibility of key metrics and understanding of operational activities. Examples of such metrics are inventory levels, days of supply, on-time-in-full (OTIF), safety stock, and lead times.
2. *Alerts:* Control towers use automated systems to identify and flag any abnormalities or deviations from the planned processes, enabling quick detection, personalized, prioritized, and prompt action.
3. *Collaboration and communication:* Facilitating collaborations across functional boundaries that exist within the supply chain is vital for control tower maturity. It involves establishing clear communication channels, facilitating information sharing, and fostering cross-functional collaboration that drives quicker, more informed decision-making.
4. *Scenario analysis:* Control towers employ advanced analytics and modeling techniques to simulate various scenarios and assess their potential impact on the supply chain. This helps in making informed decisions to mitigate risks and optimize operations.
5. *Real-time optimization/self-correction:* The most mature stage of control tower implementation involves leveraging real-time data and advanced algo-

rithms to dynamically optimize operations based on changing conditions, ensuring efficient utilization of assets and resources.

4.3.7 *Best Practices*

Several practices are highly rewarding in planning, designing, and deploying such transformation solutions. During planning, a clear vision of the objective needs to be addressed. Recommended best practice activities to be done.

- Personas: Understand the groups of users, i.e., persons, of the control tower to address their requirements jointly. Examples of such personas are groups from the following organizations in the supply chain: Make, Deliver, Plan, Finance, etc. The plan organization is often the most recognized, being the biggest beneficiary of such a program. By focusing on personas, the control tower will have a more personalized feel, which helps end users focus on what is most valuable to them and helps adoption of the system.
- Agile Practice: During all phases of projects, it is advisable to gather requirements, develop, and deploy in short scrums. Utilizing sprint demos as part of the agile methodology is an important step to get early feedback on what is built. This allows for course correction if needed, but equally important, serves as a change management opportunity because end users see their feedback incorporated and then become invested in the success of the system.
- UX (User Experience) early design: From the get-go, we need to get a clear understanding of the end in mind, stemming from actual requirements gathered from the personas. Such UX will translate into UI (User Interface). An effective interface for a control tower will focus on driving informed decision-making quickly and unambiguously.
- Change Management: Typically, there are 10 s to 1000s of users for such a tool at a large company, hence we need to make sure that requirements are well-captured, and the new way of working is vividly clear and deployable. To help dive into the new way of working, it is important to evaluate the processes that will be affected by the introduction of the control tower and the required changes to ways of working after implementation. One straightforward way to identify change is to use a current state process map and circle steps that are impacted or replaced with the introduction of the control system. This can then be incorporated into training to help users understand when and how to use the system. There is a need to get a pulse check before and after the tool is developed and utilized. Utilizing a survey or equivalent method to gather feedback can help identify the effectiveness of the tool as well as opportunities for training, procedural, or system improvements. Training material for the solution needs to be well documented and revised as the solution gets upgraded with time. The training material should be proportionate to the scope of the deployment and use multiple communication methods when possible (i.e., hands-on training,

written procedures, short video tutorials). Proper training and integration into business processes will provide the best chances for adoption of the solution. The primary objective is to ensure clarity, awareness, authenticity, execution, and sustainability throughout the entire process.

4.3.8 Special Topic: Use of Generative AI (GenAI)

On the other hand, the use of Generative Artificial Intelligence (GenAI) can be considered. The utilization of GenAI holds the potential to streamline the collection of pertinent information through a question-and-answer format, thereby expediting the decision-making process. The advantage of this approach is that it can work with largely unstructured data, with the risk of "hallucinations", which leads to incorrect data or conclusions being provided to users.

4.4 Summary and Conclusions

The significance of supply chain resilience has been brought to the forefront, especially considering the COVID-19 pandemic, which originated in 2020. Other geopolitical events have also had a considerable impact on global supply chains. For example, the sudden closure of the Suez Canal in early 2021, due to an obstruction caused by an ocean ship, led to significant delays in shipments. Add to that the reduced flow of ships in the Red Sea, forcing shipments to be shipped using the Cape of Good Hope due to a crisis in the Middle East in 2023. These collective disruptions, driven by both the pandemic and geopolitical conflicts, have underscored the urgent need for supply chains to demonstrate agility, adaptability, and resilience in the face of unexpected shocks. To tackle this challenge, businesses are increasingly exploring the potential of digital twin technology and embracing the concept of a supply chain control tower.

The adoption of digital twin technology has proven to be valuable in establishing a supply chain control tower framework, akin to real-time airport operations, but with a specific focus on managing end-to-end supply chain processes. By implementing a digital twin, businesses can create a digital replica of their supply chain, enabling them to gain real-time visibility, identify bottlenecks or disruptions, and simulate different scenarios to optimize their operations.

The construction of the supply chain control tower involves multiple layers of data acquisition, including the integration of data from siloed IT systems, data transformation, and the creation of a graph database based on the processed data. This allows for a comprehensive view of the supply chain, capturing both internal and external data sources.

In our framework, we classify digital models into four distinct types: offline, digital shadow, digital twin, and hybrid twin. The offline model relies on manual input and generates manual output based on the digital object, which is typically an algorithm using statistical or simulation methods. On the other hand, the digital shadow and digital twin models receive data automatically, with the digital shadow also generating manual output. The digital twin, being the most advanced version, automatically feeds back data into the system for adjustments. The hybrid twin combines aspects of both the shadow and digital twin models, featuring automatic data input and both manual and automatic output.

Our proposed supply chain control tower can be designed to operate as any of the three latter digital models, depending on the specific requirements and goals of the supply chain system. This flexibility allows businesses to tailor the solution to their unique needs, leveraging the benefits of real-time data analytics and decision-making.

Such a solution has become popular to manage supply chains on an operational level through four settings of business cases: visibility of network key performance indicators (KPIs) in near-real-time, alerts for potential disruptions or bottlenecks, scenario analysis platforms for simulating different scenarios, and advanced features for optimizing mitigation strategies based on previous actions.

Future trends indicate significant benefits for both commercial operations and research and development (R&D). On the commercial side, extending the control tower to gather downstream data, such as real-time points of sales (PoS) consumption data, GPS tracking data, field inventory data, and Internet of Things (IoT) sensor data, can be highly advantageous. This data provides valuable insights into product tracking, consumer behavior, demand patterns, and market trends, enabling businesses to make informed decisions and optimize their supply chain processes. Field inventory data, particularly in consignment and loan situations, can help manage assets more efficiently and ensure maximum customer satisfaction. Analyzing this data allows organizations to identify cost-saving opportunities, optimize inventory levels, and enhance overall operational efficiency. For R&D teams, leveraging control tower capabilities can significantly improve product innovation. By understanding the performance and characteristics of tested products during their initial phases, R&D teams can gain insights into yields, quality, and other critical factors. This knowledge informs the development of new and improved products, helping businesses stay competitive in the market. Additionally, the use of Design of Experiments (DoE) within control towers enables systematic experimentation and optimization of product design parameters for enhanced design for manufacturing with agility. These advancements empower businesses to drive innovation, accelerate time-to-market, and deliver products that more effectively meet customer expectations.

Overall, a supply chain control tower powered by digital twin technology offers businesses a powerful tool for managing their supply chains, mitigating risks, and optimizing operations. By harnessing the capabilities of real-time data analytics and scenario simulation, businesses can enhance their supply chain resilience and better navigate the challenges posed by both internal and external factors.

References

Aamir Rashid, R. R. (2024). Role of information processing. *Journal of Global Operations and Strategic Sourcing, 17*(2), 429–447.

Abdel-Aziz Ahmad Sharabati, S. A.-A. (2022). The use of supply chain control tower in pharmaceutical industry to create a competitive advantage. *International Journal of Pharmaceutical and Healthcare Marketing, 16*(3), 354–375.

Marmolejo-Saucedo, J. A. (2020). Design and Development of Digital Twins: a Case Study in Supply. *Mobile Networks and Applications*, 2141–2160.

Vlachos, I. (2021, Dec 13). Implementation of an intelligent supply chain control tower: a socio-technical systems case study. *Production Planning & Control, 34*(15), 1415–1431.

Part II
Enabling Technologies and Architecture

Chapter 5
The Digital Backbone of Modern Supply Chains: Enabling Technologies for Digital Twins

Bulent Soykan and Ghaith Rabadi

Abstract Modern supply chains face significant challenges related to complexity and vulnerability, necessitating advanced solutions for enhanced resilience and operational efficiency. Digital twins, serving as dynamic virtual counterparts to physical supply chain systems, present a powerful approach for real-time monitoring, sophisticated analysis, and optimized control. This chapter provides an in-depth examination of the core enabling technologies—specifically the Internet of Things (IoT), cloud computing, data analytics (including AI/ML), simulation modeling, and blockchain—that collectively form the essential digital backbone for implementing effective supply chain digital twins. We analyze the distinct role of each technology: IoT for real-time data capture, cloud computing for scalable infrastructure, data analytics for deriving predictive and prescriptive insights, simulation modeling for dynamic "what-if" analysis and optimization, and blockchain for ensuring data integrity and facilitating secure multiparty collaboration. Emphasis is placed on analyzing the crucial synergy that arises from the convergence of these technologies, demonstrating how their integration underpins the advanced capabilities required to accurately mirror, predict, and manage complex supply chain dynamics. By synthesizing insights from academic literature and referencing documented applications, the chapter illustrates the tangible benefits of leveraging these integrated technologies, including improved visibility, enhanced decision-making, optimized performance, increased resilience, and fostered trust. Ultimately, this chapter underscores the indispensable nature of this technological convergence in realizing the transformative potential of digital twins for contemporary supply chain management.

B. Soykan (✉)
Institute for Simulation and Training, University of Central Florida, Orlando, FL, USA
e-mail: Bulent.Soykan@ucf.edu

G. Rabadi
School of Modeling, Simulation and Training, University of Central Florida, Orlando, FL, USA
e-mail: Ghaith.Rabadi@ucf.edu

G. Rabadi, B. Soykan (eds.), *Optimizing Supply Chains Through Digital Twins*,
International Series in Operations Research & Management Science 375,
https://doi.org/10.1007/978-3-032-08147-6_5

Keywords Digital twin · Supply chain management · Internet of Things (IoT) · Cloud computing · Data analytics · Simulation modeling · Blockchain

5.1 Introduction to Digital Twins in Supply Chains

Recent years have witnessed a rise in the structural complexity of supply chains, alongside heightened sensitivity to disruptive forces, driven by factors such as global pandemics, geopolitical tensions, and economic volatility (Katsaliaki et al., 2022; Dolgui & Ivanov, 2021). These challenges require tools for monitoring, simulation, and optimization to ensure efficiency and resilience (Ivanov & Dolgui, 2021). Among the emerging technologies designed to meet these needs, digital twins offer a particularly powerful approach. Digital twins serve as high-fidelity virtual representations of physical assets or processes, facilitating dynamic modeling, analysis, and performance improvement (Venkatesh, 2025). In the context of supply chains, a digital twin serves as an intelligent digital counterpart to the entire supply chain network or its individual components (Bhandal et al., 2022). These virtual models are designed to accurately mirror real-world operations by continuously assimilating live data streams originating from diverse points within the system, such as machinery, sensors, and enterprise systems (Javaid et al., 2023). This integration allows digital twins to reflect the current state of physical assets processes and systems, including critical operational parameters such as machine performance, output levels, and energy consumption. Unlike traditional digital models, digital twins are dynamic and interactive, offering a holistic and current representation of supply chain activities that can be used to monitor, predict, and improve performance in real time (Mihai et al., 2022).

Achieving the level of dynamic interaction, real-time data integration, and comprehensive analysis characteristic of effective digital twins necessitates a robust technological infrastructure. The capabilities described—mirroring operations, processing vast data streams, enabling predictive insights, facilitating complex simulations, and securing transactions—do not arise spontaneously. They are fundamentally dependent on a specific suite of enabling technologies working in concert. Consequently, a deeper examination of these core technological components is crucial to fully grasp how digital twins are built, operated, and leveraged to transform supply chain management (Qi et al., 2021).

The primary objective of this chapter is to examine how key enabling technologies—specifically the Internet of Things (IoT), cloud computing, data analytics, simulation modeling, and blockchain—serve as the foundational components for creating and operating digital twins within modern supply chains. This chapter analyzes the integration of these diverse technologies, exploring how their synergy enhances the comprehensive operational capabilities of digital twins. The main research questions of this chapter include: (1) How do the key enabling technologies individually serve as foundational components for establishing and operating digital twins within the supply chain domain? (2) In what specific ways does the synergistic integration of these key enabling technologies enhance the analytical, predictive, and optimization capabilities of supply chain digital twins, thereby contributing to improved operational performance?

This chapter employs an analytical approach to investigate the synergistic relationships among IoT, cloud computing, data analytics, simulation modeling, and blockchain in the context of supply chain digital twin development. It focuses on how the convergence of these technologies provides the essential infrastructure, data flows, analytical power, modeling capabilities, and security framework required for effective implementation. To illustrate practical relevance and impact, the research incorporates findings from documented case studies and published literature demonstrating the benefits derived from integrating these technologies within digital twins in real-world supply chain scenarios.

The remainder of this chapter is organized as follows. Section 5.2 reviews common applications of digital twins within key supply chain functions such as planning, warehouse management, and transportation. Section 5.3 presents the core enabling technologies, with subsections detailing the specific roles of IoT (Sect. 5.3.1), cloud computing (Sect. 5.3.2), data analytics (Sect. 5.3.3), simulation modeling (Sect. 5.3.4), and blockchain (Sect. 5.3.5), followed by an analysis of their convergence and synergy (Sect. 5.3.6). Subsequently, Sect. 5.4 presents documented applications and benefits drawn from academic and industry sources. Finally, Sect. 5.5 concludes the chapter by summarizing the key findings and outlines potential future research directions.

5.2 Digital Twin Applications in Supply Chains

Traditional digital models in supply chain management typically offer static representations that are limited in their ability to reflect ongoing changes within the system (Sarimveis et al., 2008). These models are often built for specific scenarios or use cases and lack the flexibility to evolve as the supply chain dynamics shift. In contrast, digital twins provide a continuously updating virtual counterpart that evolves in tandem with the physical supply chain (Kamble et al., 2022). This dynamic nature allows digital twins to not only represent the current state of the supply chain accurately but also to anticipate and adapt to future changes. Unlike traditional models, which require substantial effort to update and recalibrate, digital twins autonomously incorporate new data streams, ensuring that the virtual representation remains relevant and precise over time.

Another key distinction lies in the level of detail and fidelity between digital twins and traditional digital models (Tao et al., 2022; Wright & Davidson, 2020). Traditional models may oversimplify complex supply chain processes to make simulations manageable, potentially overlooking critical nuances that can impact decision-making. Digital twins, however, leverage high-resolution data from a multitude of sources, including IoT sensors, radio frequency identification (RFID) tags, and real-time tracking systems, to create highly detailed and accurate representations of each component within the supply chain (Le et al., 2024). This granular level of detail enables more precise monitoring, analysis, and optimization, allowing organizations to identify and address inefficiencies that might remain hidden in less detailed models. The enhanced fidelity of digital twins thus facilitates a deeper understanding of supply chain operations and supports more informed strategic decisions.

In addition, digital twins integrate advanced analytics and machine learning (ML) algorithms to provide predictive and prescriptive insights that surpass the capabilities of traditional digital models (Al Zami et al., 2025). While traditional models may primarily focus on descriptive analytics—summarizing past and present states—digital twins utilize predictive analytics to forecast future trends and potential disruptions. By analyzing historical data alongside real-time information, digital twins can anticipate demand fluctuations, identify potential risks, and suggest optimal responses (Mhaskey, 2024). Also, prescriptive analytics enable digital twins to recommend specific actions to optimize supply chain performance, such as adjusting inventory levels or rerouting transportation. This proactive approach empowers supply chain managers to make data-driven decisions that enhance resilience and efficiency, moving beyond the reactive nature of traditional modeling approaches (Owusu-Berko, 2025).

Interactivity and user engagement represent another area where digital twins differentiate themselves from traditional digital models (Yigitbas et al., 2021). Traditional models generally serve as passive tools, providing static outputs that require manual interpretation and action by users. In contrast, digital twins offer interactive interfaces that allow users to engage with the virtual model in real time. Supply chain managers can manipulate variables, run simulations, and visualize outcomes directly within the digital twin environment. This interactive capability fosters a more intuitive and hands-on approach to supply chain management, enabling users to experiment with different scenarios and immediately observe the potential impacts of their decisions. The enhanced interactivity not only improves user experience but also accelerates the decision-making process by providing immediate feedback and insights.

Scalability and flexibility are also significant differentiators of digital twins compared to traditional digital models (Jiang et al., 2021). Traditional models are often constrained by their initial design parameters and may struggle to scale effectively as the supply chain grows or becomes more complex. Expanding or modifying these models typically requires substantial redevelopment and can be time-consuming and costly. Digital twins, on the other hand, are built on scalable architectures that can easily accommodate additional data sources, expanded operational scopes, and increased complexity without necessitating major overhauls (Soykan et al., 2025). This scalability ensures that digital twins remain effective and relevant even as supply chains evolve, allowing organizations to adapt swiftly to growth, diversification, or changes in market conditions. The inherent flexibility of digital twins thus supports long-term sustainability and continuous improvement within supply chain operations.

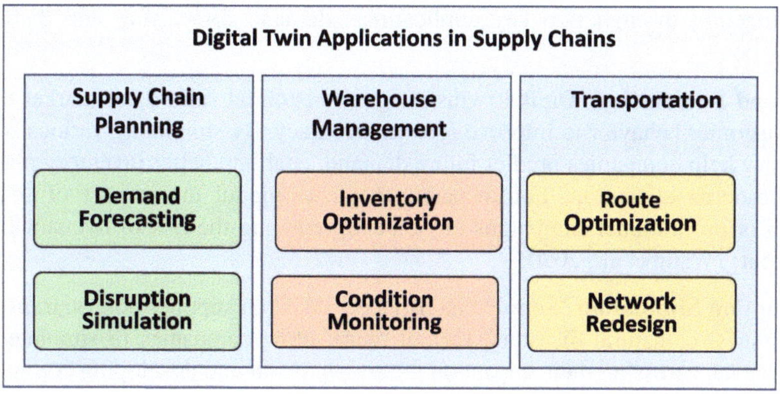

Fig. 5.1 Digital twin applications in supply chains

Lastly, the integration capabilities of digital twins with other advanced technologies set them apart from traditional digital models (Erkoyuncu et al., 2018). Digital twins can incorporate inputs from AI, ML, blockchain, and other emerging technologies to enhance their functionality and value. For example, AI and ML algorithms can be embedded within digital twins to enable autonomous decision-making, anomaly detection, and optimization without human intervention (Rojas et al., 2025). Also, digital twins can interface with enterprise resource planning (ERP) systems, customer relationship management (CRM) tools, and other business intelligence platforms to create a unified and comprehensive view of the supply chain (Sarker, 2023). This interoperability not only enhances the overall utility of digital twins but also ensures that they can support a wide range of business processes and strategic initiatives, making them indispensable tools for modern supply chain management.

Using these significant advantages, including dynamic update, high fidelity, advanced analytics, interactivity, scalability, and integration capabilities, digital twins are being applied across various critical functions within supply chain management. These applications translate theoretical benefits into practical tools to optimize performance, improve resilience, and improve decision-making. Figure 5.1 illustrates several prominent application areas, namely supply chain planning, warehouse management, and transportation, along with specific use cases within each domain. The following subsections will present these applications in detail.

5.2.1 Digital Twins in Supply Chain Planning

Supply chain planning involves forecasting demand, managing inventory, and preparing for disruptions to ensure efficient operations. Digital twins enhance this function through two key applications: demand forecasting and disruption simulation.

Demand Forecasting Digital twins integrate historical sales data, market trends, and customer behavior to improve forecast accuracy. By simulating various scenarios, they help companies predict future demand, enabling better inventory planning and resource allocation. For instance, they can model the impact of seasonal trends or promotional campaigns on demand, reducing the risk of overstocking or stockouts (Wang et al., 2020).

Disruption Simulation Supply chains face risks from supplier delays, transportation issues, or natural disasters. Digital twins allow companies to simulate these disruptions, assessing their impact on the supply chain and developing contingency plans. This proactive approach minimizes risks, ensuring operational continuity and cost efficiency (Ivanov et al., 2019).

5.2.2 Digital Twins in Warehouse Management

Warehouse management focuses on maintaining inventory accuracy, optimizing space, and ensuring efficient order fulfillment. Digital twins in this domain emphasize inventory optimization and condition monitoring, enhancing operational efficiency and reducing downtime.

Inventory Optimization Digital twins enable continuous monitoring of inventory quantities, asset positions, and material flow, integrating data from warehouse management systems (WMS) and IoT sensors. This enables companies to optimize inventory levels, reduce carrying costs, and prevent stockouts or overstocking. They can simulate different inventory strategies, such as just-in-time delivery, to find the most efficient approach (Roca Zúñiga, 2023).

Condition Monitoring For perishable or sensitive goods, maintaining optimal storage conditions is critical. Digital twins monitor environmental parameters like temperature, humidity, and light, alerting managers to deviations that could affect product quality. This is particularly valuable in industries like pharmaceuticals, food, and electronics, where condition breaches can lead to significant losses (Wang et al., 2024; Maheshwari et al., 2023).

5.2.3 Digital Twins in Transportation

Transportation involves the efficient movement of goods, requiring optimized routes, schedules, and network designs. Digital twins enhance this function through route optimization and network redesign, lowering costs and improving responsiveness.

Table 5.1 Digital twin application in supply chain and its impact on efficiency

Supply chain function	Digital twin application	Impact on efficiency
Supply chain planning	Demand forecasting, disruption simulation	Improves accuracy, reduces risk
Warehouse management	Inventory optimization, condition monitoring	Enhances just-in-time delivery, reduces downtime
Transportation	Route optimization, network redesign	Lowers costs, improves responsiveness

Route Optimization Digital twins analyze factors such as traffic conditions, weather, vehicle availability, and delivery windows to determine the most efficient routes. By simulating different routing scenarios, companies can minimize transportation costs, reduce delivery times, and improve customer satisfaction. They can also account for dynamic changes, such as road closures or congestion, ensuring timely deliveries (Tozanli & Saénz, 2022).

Network Redesign For companies with complex logistics networks, digital twins model the entire transportation system, including suppliers, manufacturing sites, warehouses, and distribution centers. This holistic view allows for the redesign of networks to optimize flow, reduce bottlenecks, and enhance overall efficiency. Simulations can test the impact of adding new routes, changing carriers, or consolidating shipments, supporting strategic decision-making (Lim et al., 2024).

To illustrate the breadth of digital twin applications, Table 5.1 compares key use cases across different supply chain functions, highlighting their impact on efficiency and decision-making. This table underscores the versatility of digital twins, addressing specific needs within each function to drive overall supply chain performance. While supply chain planning focuses on strategic forecasting and risk management, warehouse management emphasizes operational efficiency through real-time tracking, and transportation management prioritizes logistical optimization for cost and responsiveness. Commonalities include reliance on real-time data integration and advanced analytics, with differences in the scale and scope of application, from tactical inventory decisions to strategic network redesigns.

5.3 Enabling Technologies for Digital Twins in Supply Chains

The capabilities of supply chain digital twins, ranging from real-time mirroring to predictive analysis and automated optimization, are not inherent properties but are realized through the integration of several key enabling technologies (Zhang et al., 2022). These technologies form the foundational infrastructure and provide the essential functions required to capture data from the physical world, store and process it efficiently, derive meaningful insights, model complex behaviors, and ensure secure interactions across the supply network. Understanding the specific roles and synergistic interplay of these core components is crucial for appreciating how digital twins function and deliver value.

5.3.1 *Internet of Things (IoT)*

IoT is fundamental to digital twins as it provides the real-time data necessary to keep the virtual model synchronized with its physical counterpart (Minerva et al., 2020). IoT devices, such as sensors, RFID tags, and smart meters, are embedded in physical assets throughout the supply chain. These devices continuously collect data on various parameters, including location, temperature, humidity, machine performance, and more. For example, in warehouse management, IoT sensors can monitor inventory levels and environmental conditions, ensuring that products are stored optimally. In transportation, IoT devices track vehicle locations and conditions, providing real-time visibility into logistics operations. This data is then transmitted to the digital twin, allowing it to reflect the current state of the physical system accurately. A systematic review by Zaidi et al. (2024) highlights IoT and RFID as key technologies, with simulation modeling being the most widely used data analytics technique, emphasizing their role in enhancing supply chain performance through real-time connectivity and data-driven modeling.

IoT serves as the fundamental sensory layer for supply chain digital twins, acting as the crucial bridge between physical operations and their digital representation. IoT encompasses a network of interconnected physical objects—ranging from individual packages and containers to manufacturing equipment, vehicles, and warehouse infrastructure—embedded with sensors, actuators, software, and connectivity (Muhic & Hodzic, 2014). Its primary role in the context of digital twins is the continuous, automated capture and transmission of real-time data reflecting the status, condition, location, and performance of assets and processes throughout the supply chain. Without this constant stream of high-fidelity data from the physical world, a digital twin would remain a static model, unable to mirror the dynamic nature of actual operations.

IoT devices capture a diverse array of data critical for maintaining an accurate digital twin. Key examples within the supply chain include:

- **Asset Tracking and Condition Monitoring:** GPS sensors provide real-time location data for goods in transit, while environmental sensors monitor parameters like temperature, humidity, shock, and tilt for sensitive products, ensuring compliance and quality (Rehman et al., 2024). RFID tags and smart labels enable automated identification and tracking as items move through different nodes.
- **Manufacturing and Production:** Sensors embedded in machinery monitor operational status (on/off, speed, load), output levels, energy consumption, temperature, vibration, and other indicators vital for assessing performance, pre-

dicting maintenance needs (predictive maintenance), and optimizing production schedules (Zanella et al., 2014).
- **Logistics and Transportation:** Telematics systems in vehicles capture data on location, speed, fuel efficiency, engine diagnostics, and driver behavior. Sensors on cargo can verify load integrity and security.
- **Warehouse and Inventory Management:** Smart shelves, weight sensors, and vision systems can provide real-time inventory levels, while environmental sensors monitor storage conditions. Automated guided vehicles (AGVs) and drones used in warehouses also generate operational data.

The capture mechanisms involve various sensor types (e.g., thermal, optical, acoustic, motion, chemical) coupled with identification technologies (RFID, NFC, QR codes) and diverse communication protocols (e.g., Wi-Fi, Bluetooth, Cellular (4G/5G)) suited to different ranges, power requirements, and data volumes (Al-Fuqaha et al., 2015). This data is then transmitted, often via edge gateways, to backend systems—typically the cloud platforms hosting the digital twin—for processing and integration.

The significance of IoT-derived data for supply chain digital twins cannot be overstated. It provides the empirical grounding that allows the twin to accurately reflect the current state of the physical system. This real-time visibility enables immediate detection of deviations, anomalies, or potential disruptions, facilitating proactive responses rather than reactive corrections. Furthermore, this continuous data feed is the essential input for the analytical and simulation components of the digital twin, allowing for informed decision-making, performance optimization, and reliable "what-if" scenario analysis based on actual operating conditions. In essence, IoT acts as the distributed nervous system of the physical supply chain, constantly feeding the digital brain—the twin—with the information needed to understand, manage, and improve operations.

5.3.2 Cloud Computing

Cloud computing offers the scalable infrastructure required to store and process the vast amounts of data generated by IoT devices (Stergiou et al., 2018). It provides the computing power necessary to run complex simulations and analyses that are integral to digital twins. Cloud platforms enable the storage of historical data, which, when combined with real-time data, allows for comprehensive analysis and forecasting. Moreover, cloud computing facilitates collaboration and data sharing across different stakeholders in the supply chain, ensuring that all parties have access to the same up-to-date information. Cloud computing has fundamentally altered IT cost structures, facilitating on-demand access to extensive computational power and allowing resource allocation to dynamically align with requirements, which is crucial for handling the data volumes in digital twins (Zhang et al., 2022).

While IoT provides the data streams from the physical world, cloud computing delivers the essential, scalable, and flexible infrastructure required to host, process, and manage the vast amounts of information and complex models that constitute a supply chain digital twin (Marston et al., 2011). The sheer volume, velocity, and variety of data generated by IoT sensors, combined with data from enterprise systems (like ERP, WMS, TMS) and the computational demands of simulation and analytics, quickly exceed the capabilities of traditional on-premise IT infrastructure. Cloud platforms offer the necessary backend power and flexibility, acting as the central nervous system's processing center.

A primary contribution of cloud computing is providing virtually limitless and elastic storage capacity. Supply chain digital twins accumulate massive datasets, including continuous real-time feeds from IoT devices, extensive historical performance data, complex 3D models, simulation parameters, and analytical results. Cloud storage solutions (e.g., object storage, data lakes, managed databases) allow organizations to store this data cost-effectively and scale their storage needs up or down based on demand, without significant upfront investment in physical hardware (Hashem et al., 2015). This elasticity ensures that data storage never becomes a bottleneck as the digital twin evolves or as data generation rates fluctuate.

Equally critical is the provision of on-demand, scalable computational power. Running sophisticated digital twin models, performing complex simulations for "what-if" scenario analysis, executing advanced analytics algorithms (including ML and AI for predictive insights), and rendering visualizations require substantial processing resources. Cloud providers offer access to a wide range of computing options, from standard virtual machines to powerful GPUs (graphics processing units) optimized for parallel processing and AI workloads, as well as serverless computing functions (Armbrust et al., 2010). Crucially, this computational power is elastic; resources can be dynamically allocated to match the immediate needs of the digital twin—scaling up for intensive simulation runs or large-scale data analysis, and scaling down during periods of lower activity. This pay-as-you-go model optimizes costs and ensures performance is maintained even under heavy load.

Cloud platforms also enhance accessibility and collaboration. By hosting the digital twin in the cloud, authorized stakeholders across geographically dispersed locations and different functional departments (e.g., planning, logistics, manufacturing, procurement) can access the twin's data, models, and insights through web interfaces or APIs (Buyya et al., 2009). This centralized accessibility fosters better cross-functional collaboration and ensures decision-makers are working with a consistent, up-to-date view of the supply chain. Cloud environments also often provide integrated development, deployment, and management tools (DevOps, MLOps) that streamline the lifecycle management of the digital twin application itself.

Cloud computing serves as the indispensable infrastructural foundation upon which viable and powerful supply chain digital twins are built. Its inherent scalability, elasticity, accessibility, and managed service offerings allow organizations to handle the immense data and computational requirements effectively and economically, making the sophisticated capabilities of digital twins practically achievable.

5.3.3 Data Analytics

Data analytics, encompassing techniques such as AI, ML, and predictive modeling, is crucial for extracting actionable insights from the data collected. These analytics tools process both real-time and historical data to identify patterns, predict future states, and optimize operations. In supply chain planning, data analytics can enhance demand forecasting by analyzing sales history, market trends, and customer behavior. In warehouse management, analytics can optimize inventory levels and predict equipment failures, reducing downtime. Hybrid data analytics, combining simulation with AI and ML, is proposed for decentralized decision support, increasing visibility for real-time tracking, better risk management, and customized supply chains based on market demand (Zaidi et al., 2024).

While IoT provides the raw data and Cloud Computing offers the infrastructure, Data Analytics represents the core intelligence engine within the supply chain digital twin. Its fundamental purpose is to process, analyze, and interpret the vast and complex datasets collected from various sources—including IoT sensors, enterprise systems, historical records, and even external sources like weather or market data—to extract meaningful insights, generate accurate predictions, and ultimately drive informed decision-making and optimization (Chen et al., 2012). Without analytics, the digital twin would merely be a passive repository of data rather than an active tool for performance enhancement.

The analytics applied within a supply chain digital twin typically span several levels of increasing complexity and value:

- **Descriptive Analytics:** Answers the question "What is happening?" This involves processing real-time data streams to provide dashboards, reports, and key performance indicators (KPIs) that offer immediate visibility into the current state of operations, such as inventory levels, production output, shipment locations, and equipment status (Sharda et al., 2014).
- **Diagnostic Analytics:** Seeks to understand "Why did it happen?" By drilling down into historical and real-time data, diagnostic analytics helps identify the root causes of problems or deviations, such as determining the factors leading to a production bottleneck or a delayed shipment (Wolniak & Grebski, 2023).
- **Predictive Analytics:** Addresses "What is likely to happen?" Leveraging historical data patterns and advanced statistical modeling, ML, and AI techniques, this level forecasts future events or trends. Examples include predicting equipment failures (predictive maintenance), forecasting customer demand, estimating shipment arrival times, and identifying potential disruption risks (Waller & Fawcett, 2013).

- **Prescriptive Analytics:** Focuses on "What should we do about it?" This is the most advanced level, using optimization algorithms, simulation results (often linked with the simulation modeling component), and AI to recommend specific actions to achieve desired outcomes or mitigate predicted problems. Examples include suggesting optimal inventory replenishment levels, recommending alternative transportation routes to avoid disruptions, or automatically adjusting machine parameters for better efficiency (Lepenioti et al., 2020).

The implementation of these analytics relies heavily on the underlying cloud infrastructure for the necessary computational power to train complex ML models and process large datasets rapidly. Techniques ranging from statistical process control and time-series analysis to deep learning and reinforcement learning may be employed, depending on the specific application (Soykan & Rabadi, 2024). The outputs are often visualized within the digital twin interface, providing intuitive insights to users, or used to trigger automated alerts and actions within connected systems.

5.3.4 Simulation Modeling

While data analytics provides insights into past and present performance and predicts future trends, simulation modeling within a supply chain digital twin offers a powerful capability to dynamically experiment with the system under various hypothetical conditions (Soykan & Rabadi, 2024). It involves creating executable models that replicate the behavior and interactions of supply chain components and processes over time. Integrated within the digital twin framework, these simulation models use the real-time data feeds and the comprehensive digital representation to provide a virtual testbed for evaluating strategies, optimizing processes, and assessing risks without impacting physical operations.

A core function of simulation modeling in digital twins is conducting "what-if" scenario analysis. Managers can use the simulation model, initialized with current or projected states derived from the twin's data, to explore the potential impacts of various decisions or external events (Soykan et al., 2023). For instance, they can simulate:

- The effects of changing inventory policies or sourcing strategies
- The consequences of potential disruptions, such as a supplier outage, port congestion, or transportation lane closure
- The impact of introducing new technologies, equipment, or processes within a facility
- The supply chain's ability to respond to sudden shifts in customer demand or market conditions

By running these scenarios virtually, organizations can gain foresight into potential outcomes and make more informed, proactive decisions.

Beyond exploratory analysis, simulation modeling is a key enabler of process optimization and risk mitigation. Different operational parameters, layouts, resource allocations, or control logic can be tested within the simulation environment to identify configurations that maximize efficiency, minimize costs, improve throughput, or enhance resilience. Techniques such as discrete-event simulation (DES) are commonly employed to model queuing systems, resource contention, and process flows typical in manufacturing and logistics (Banks, 2005). Agent-based modeling (ABM) can also be used to capture the complex interactions and emergent behaviors of autonomous entities within the supply chain, such as vehicles, suppliers, or customers (Macal & North, 2014). The results of these experiments can inform prescriptive analytics, helping to select the best course of action among simulated alternatives. Also, simulating various failure modes and disruption scenarios allows for quantitative risk assessment and the validation of contingency plans.

The fidelity and reliability of these simulations are heavily dependent on the accuracy and timeliness of the data feeding into the digital twin from IoT devices and enterprise systems, as well as the validity of the underlying simulation model logic. The computational demands of running numerous complex simulation scenarios often necessitate leveraging the scalable processing power offered by the cloud computing infrastructure.

5.3.5 *Blockchain*

In complex, multi-enterprise supply chains, establishing trust and enabling secure, transparent data sharing among numerous independent partners are significant challenges. Blockchain technology offers a potential solution by providing a distributed, immutable, and transparent ledger system that can significantly enhance the reliability and collaborative potential of supply chain digital twins (Kshetri, 2018). While other technologies focus on data capture, storage, analysis, and modeling, blockchain primarily addresses the critical aspects of data integrity, provenance, and secure multiparty interactions within the digital twin ecosystem.

The core contribution of blockchain stems from its inherent properties. Transactions or data entries are cryptographically linked into blocks, forming a chain that is replicated across multiple nodes in a distributed network. This architecture provides:

- **Immutability:** Once data is recorded on the blockchain, it becomes extremely difficult and computationally expensive to alter or delete it without detection, ensuring a high degree of tamper resistance.
- **Transparency and Provenance:** Depending on the blockchain type (public, private, consortium), authorized participants can view the recorded transactions, providing a shared, auditable history of events or data exchanges. This facilitates tracking the origin and journey of goods or information (provenance).

- **Decentralization:** Control and data validation are distributed among network participants rather than residing with a single central authority, reducing single points of failure and potentially mitigating issues of data ownership and control bias (Swan, 2015).

Applied to supply chain digital twins, blockchain can significantly bolster data integrity. Data points originating from IoT sensors or entered by different supply chain partners (e.g., quality checks, customs clearance, handover events) can be recorded on a blockchain. This ensures that the data feeding into the digital twin, particularly data shared across organizational boundaries, is trustworthy and has not been tampered with (Saberi et al., 2019). An immutable record of critical events enhances the reliability of the digital twin's representation of the physical supply chain's state and history.

Blockchain also facilitates secure multiparty collaboration. It provides a shared, trusted infrastructure where different organizations (suppliers, manufacturers, logistics providers, retailers, regulators) can selectively share relevant data points required for an end-to-end digital twin view, without necessarily exposing all their sensitive internal data. Permissioned blockchains allow control over who can access and contribute data (Casino et al., 2019). Moreover, smart contracts—self-executing contracts with the terms of the agreement directly written into code—can be deployed on the blockchain. These can automate processes based on verified data recorded on the ledger, such as triggering payments upon confirmed delivery or automatically updating the digital twin's status when specific conditions are met according to agreed-upon rules (Wang et al., 2019).

While challenges related to scalability, interoperability with legacy systems, energy consumption (for some blockchain types), and governance models need careful consideration (Bahga & Madisetti, 2016), blockchain's potential to establish a trusted data layer is compelling. It complements the other enabling technologies by providing a mechanism to enhance the veracity of the data being sensed, stored, analyzed, and simulated within the digital twin, thereby fostering greater confidence and enabling more robust, secure collaboration across the extended supply chain network.

5.3.6 Synergy of Enabling Technologies

While the preceding sections have detailed the individual contributions of IoT, cloud computing, data analytics, simulation modeling, and blockchain, the true transformative potential for supply chain digital twins emerges from the synergistic convergence of these technologies. It is their integration and interplay that creates a holistic system far more capable than the sum of its individual parts, forming the robust digital backbone required for sophisticated digital mirroring and management.

The relationship between these technologies is inherently interconnected and mutually reinforcing:

- *IoT* acts as the primary data source, feeding real-time observations from the physical supply chain into the digital realm.
- *Cloud computing* provides the indispensable scalable infrastructure, housing the massive datasets generated by IoT and other sources, and supplying the elastic computational power needed for *data analytics* and *simulation modeling.*
- *Data analytics* processes and refines the raw data stored in the cloud, extracting crucial insights, identifying patterns, generating predictions, and informing the parameters used within simulation models.
- *Simulation modeling* leverages the insights from analytics and the computational resources of the cloud to create dynamic models, allowing for "what-if" analysis and optimization based on realistic, data-driven scenarios.
- *Blockchain* underpins the ecosystem by enhancing the integrity and trustworthiness of the data shared among partners (often originating from IoT or manual inputs) and facilitating secure, automated interactions via smart contracts, potentially triggered by insights from analytics or simulation outcomes.

This convergence creates powerful synergistic effects. For example, the combination of real-time IoT data streams with scalable cloud processing enables analytics engines to provide near-instantaneous visibility and diagnostic capabilities. Adding simulation modeling allows these real-time insights to be projected forward, enabling predictive and prescriptive capabilities through dynamic scenario testing. Also, integrating blockchain ensures that the data flowing through this entire process, especially across organizational boundaries, is reliable and tamper-evident, increasing confidence in the digital twin's outputs and the decisions derived from them (Min, 2019). The ability to run complex simulations (simulation modeling) based on high-volume, real-time data (IoT), processed by advanced algorithms (data analytics) on powerful infrastructure (cloud), with cross-party data trust enhanced by distributed ledgers (blockchain), exemplifies this synergy.

To visualize this intricate interplay, Fig. 5.2 depicts the integrated technology stack that forms the foundation of a supply chain digital twin. It illustrates the flow of data from the physical supply chain through the various technological layers and back, highlighting the specific roles and dependencies among the core enabling components.

Consider the analogy of a biological organism: IoT acts as the sensory organs, perceiving the external environment. The cloud represents the scalable neural infrastructure and memory capacity. Data analytics serves as the cognitive processing function, interpreting sensory input and recognizing patterns. Simulation modeling functions like the capacity for imagination and planning, exploring potential futures based on current understanding. Blockchain acts akin to a secure immune system or a trusted memory verification mechanism, ensuring the integrity of information and interactions. Only through the integrated functioning of all these systems can the organism (the supply chain, managed via its digital twin) perceive, understand, predict, plan, and act effectively and securely in a complex environment.

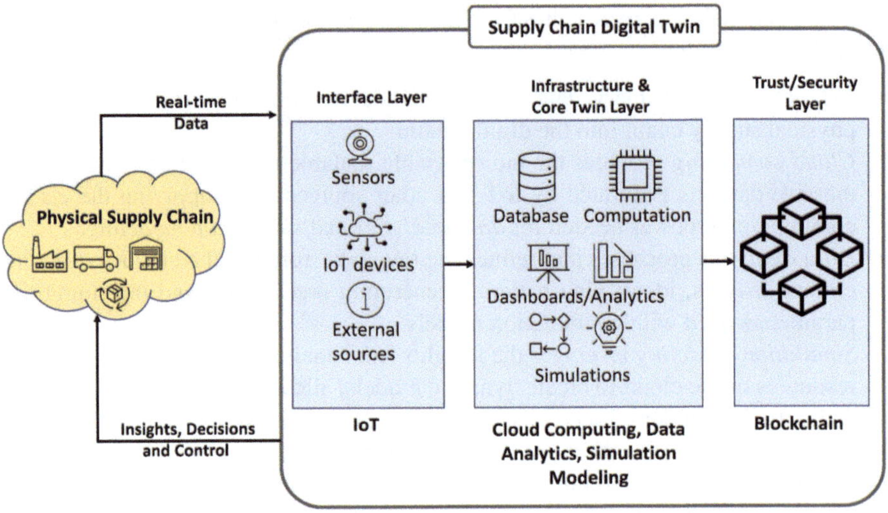

Fig. 5.2 Enabling technologies for supply chain digital twins. The figure shows the flow from the physical supply chain via IoT data capture to the cloud platform, which provides storage and computation for data analytics and simulation modeling. Blockchain underpins data integrity and trust. Insights derived lead to actions that optimize the physical system, closing the loop

5.4 Documented Applications and Benefits

The practical value of supply chain digital twins, enabled by the convergence of technologies discussed previously, is increasingly documented in academic literature and industry reports. This section reviews specific application areas where these integrated systems have demonstrated tangible benefits, drawing upon published findings and examples.

5.4.1 Enhancing Predictive Maintenance in Manufacturing Operations

Predictive maintenance (PdM) represents a prominent application where digital twins deliver significant operational improvements by forecasting equipment failures before they occur. Implementations typically leverage IoT sensors (monitoring vibration, temperature, acoustics, etc.) to stream real-time condition data to cloud platforms. Advanced analytics, particularly machine learning algorithms, then process this data alongside historical records to predict failure probabilities and remaining useful life (RUL) (Aivaliotis et al., 2019).

Simulation models within the twin can further test the impact of different maintenance interventions based on these predictions. Documented benefits from such implementations include significant reductions in unplanned machine downtime, optimized maintenance scheduling leading to lower costs, and extended asset lifespan (Werner et al., 2019). For instance, specific applications in aerospace manufacturing have shown notable improvements in component reliability through digital twin-based monitoring (Li et al., 2021).

5.4.2 Improving Cold Chain Integrity and Pharmaceutical Traceability

Maintaining the integrity of temperature-sensitive supply chains (cold chains), particularly for pharmaceuticals and perishable foods, is critical for safety and compliance. Digital twins integrating IoT sensors (for real-time temperature, humidity, location tracking) provide continuous visibility. Data fed to cloud platforms allows analytics engines to monitor for excursions from required conditions and trigger immediate alerts (Shashi, 2022). Several studies highlight the role of blockchain in this context, providing an immutable and auditable ledger for temperature records and chain-of-custody information, accessible to relevant stakeholders (shippers, regulators, recipients) (Kamble et al., 2020; Hulea et al., 2018). This enhances traceability and simplifies compliance verification (e.g., with FDA regulations). Reported outcomes include reductions in product spoilage, improved regulatory compliance rates, and enhanced trust among supply chain partners due to transparent, verifiable data trails (Panigrahi et al., 2024).

5.4.3 Boosting End-to-End Supply Chain Visibility and Resilience

Creating a holistic digital twin of the extended supply chain network addresses challenges related to visibility and disruption management. This involves integrating data from diverse sources—including IoT tracking of goods, partner systems (via APIs), ERP/TMS/WMS data—onto a unified cloud platform (Ivanov et al., 2022). Advanced analytics applied to this comprehensive dataset enable better demand forecasting, inventory optimization across multiple echelons, and early identification of potential bottlenecks or risks (Ogunsoto et al., 2025). Simulation modeling plays a crucial role here, allowing organizations to conduct "what-if" analyses on the digital twin to assess the impact of potential disruptions (e.g., supplier failures, transportation delays, geopolitical events) and evaluate the effectiveness of different mitigation strategies (e.g., multisourcing, strategic buffering) (Nguyen et al., 2023). Studies analyzing network-level digital twins report benefits such as improved responsiveness to demand changes, reduced overall inventory holding costs, faster recovery times from disruptions, and enhanced strategic decision-making regarding network design and risk management policies (Rinaldi et al., 2025).

While these examples highlight significant potential, the reported success often depends on factors such as data quality, integration maturity, organizational readiness, and the specific context of the implementation (Bakhshi et al., 2024; Sandén & Falk, 2021). Nonetheless, the documented evidence confirms that the synergistic application of enabling technologies within a digital twin framework provides powerful tools for enhancing supply chain performance across various critical functions.

5.5 Conclusion and Future Research Directions

In response to the increasing complexities and vulnerabilities facing modern supply chains, digital twins have emerged as a transformative paradigm, offering dynamic virtual representations for enhanced monitoring, analysis, and optimization. This chapter has argued that the successful realization and operation of these sophisticated digital counterparts are fundamentally dependent upon a cohesive ecosystem of enabling technologies. We have explored the distinct, yet complementary roles of the enabling technologies as the essential components constituting the digital backbone of supply chain digital twins.

Our analysis detailed how IoT acts as the sensory layer capturing real-time data, cloud computing provides the necessary scalable infrastructure for storage and computation, data analytics extracts vital intelligence and predictive insights, simulation modeling enables dynamic "what-if" analysis and optimization, and blockchain ensures data integrity and secure multiparty collaboration. Crucially, we emphasized that it is the synergistic convergence of these five technologies that unlocks the full potential of digital twins. Their integration facilitates the flow from data acquisition to actionable insight, enabling unprecedented levels of visibility, predictive capability, process optimization, and operational resilience across the supply network. The practical examples serve to illustrate the benefits achievable through this integrated technological approach.

While the potential is immense, the implementation of comprehensive digital twins underpinned by these technologies is not without challenges. Issues related to system integration, data interoperability, cybersecurity, the cost of deployment, and the need for specialized skills require careful consideration by organizations embarking on this journey. Also, future advancements, particularly in areas such as edge computing (complementing cloud), more sophisticated AI/ML algorithms within analytics, standardization of protocols, and enhanced cybersecurity measures for interconnected systems, will continue to shape the evolution and capabilities of supply chain digital twins.

While the convergence of the enabling technologies provides a powerful foundation for supply chain digital twins, significant opportunities and challenges remain, paving the way for future research. Some of the promising directions for advancing the theory, technology, and application of digital twins in supply chain management include:

- **Enhanced Sensing and Connectivity (IoT):** Research is needed on developing more cost-effective, energy-efficient, and miniaturized sensors capable of capturing a wider range of granular data (e.g., product quality degradation, micro-environmental conditions). Also, exploring the implications and integration challenges of next-generation communication networks (e.g., 6G) for ultra-low latency and massive device connectivity within the digital twin ecosystem is important. Security protocols specifically designed for large-scale, heterogeneous IoT deployments in supply chains also need further investigation.
- **Optimized Computing Architectures (Cloud/Edge):** Future work should investigate optimal hybrid cloud-edge computing architectures tailored for supply chain digital twins. This includes determining which data processing, analytics, and simulation tasks are best performed at the edge versus in the cloud to balance latency, bandwidth, cost, and security requirements. Research into federated learning approaches to train AI models across distributed data sources without compromising data privacy is also a key area.
- **Advanced Intelligence and Autonomy (Analytics/AI):** There is significant scope for research into more sophisticated AI/ML techniques within digital twins. This includes developing more accurate and robust predictive models, exploring explainable AI (XAI) methods to enhance trust and transparency in decision support, and investigating reinforcement learning for enabling autonomous control loops where the digital twin can directly optimize physical operations based on real-time data and simulations. Integrating unstructured data sources (e.g., news feeds, social media sentiment, weather patterns) more effectively into predictive models is another promising research area.
- **High-Fidelity and Real-Time Simulation (Modeling):** Advancing simulation modeling involves research into automated model generation and calibration techniques based on real-time data feeds. Integrating different simulation paradigms (e.g., discrete-event, agent-based, system dynamics) within a single digital twin framework presents challenges and opportunities. Further research is needed on achieving near real-time synchronization between the simulation model and the physical system, enabling more dynamic and responsive analysis and control.
- **Scalable and Interoperable Trust Mechanisms (Blockchain):** Addressing the scalability limitations of current blockchain technologies for high-throughput supply chain applications remains a critical research area. Developing standards and protocols for interoperability between different blockchain platforms and legacy systems is essential for wider adoption. Research into efficient and fair governance models for consortium blockchains used in supply chain ecosystems, along with lightweight consensus mechanisms suitable for IoT integration, is also needed.
- **Holistic Integration and Cybersecurity:** Frameworks and standardized architectures for integrating these diverse technologies are vital. A major cross-cutting research theme is end-to-end cybersecurity across the entire technology stack, from securing individual IoT devices to protecting cloud infrastructure, analytical models, simulation environments, and blockchain interactions.

- **Human-Twin Interaction and Socio-Technical Aspects:** Research should explore effective human-computer interaction designs for complex digital twins, ensuring usability and facilitating trust in AI-driven recommendations. Understanding the organizational changes, skill requirements, and ethical considerations associated with deploying and managing sophisticated digital twins is also crucial.
- **Sustainability and Circular Economy Applications:** Investigating how digital twins, enabled by these technologies, can be specifically designed and utilized to optimize resource usage, minimize waste, track carbon footprints, and support circular economy models within supply chains represents a significant area for future impactful research.

Addressing these research directions will be instrumental in overcoming current limitations and unlocking the next wave of innovation in supply chain digital twins.

References

Aivaliotis, P., Georgoulias, K., & Chryssolouris, G. (2019). The use of digital twin for predictive maintenance in manufacturing. *International Journal of Computer Integrated Manufacturing, 32*(11), 1067–1080.

Al-Fuqaha, A., Guizani, M., Mohammadi, M., Aledhari, M., & Ayyash, M. (2015). Internet of things: A survey on enabling technologies, protocols, and applications. *IEEE Communications Surveys & Tutorials, 17*(4), 2347–2376.

Al Zami, M. B., Shaon, S., Quy, V. K., & Nguyen, D. C. (2025). Digital twin in industries: A comprehensive survey. *IEEE Access, 13*, 47291–47336. https://doi.org/10.1109/ACCESS.2025.3551532

Armbrust, M., Fox, A., Griffith, R., Joseph, A.D., Katz, R., Konwinski, A., Lee, G., Patterson, D., Rabkin, A., Stoica, I., et al. (2010). A view of cloud computing. *Communications of the ACM, 53*(4), 50–58.

Bahga, A., & Madisetti, V.K. (2016). Blockchain platform for industrial internet of things. *Journal of Software Engineering and Applications, 9*(10), 533–546.

Bakhshi, S., Ghaffarianhoseini, A., Ghaffarianhoseini, A., Najafi, M., Rahimian, F., Park, C., & Lee, D. (2024). Digital twin applications for overcoming construction supply chain challenges. *Automation in Construction, 167*, 105679.

Banks, J. (2005). *Discrete event system simulation*. Pearson Education India.

Bhandal, R., Meriton, R., Kavanagh, R.E., & Brown, A. (2022). The application of digital twin technology in operations and supply chain management: A bibliometric review. *Supply Chain Management: An International Journal, 27*(2), 182–206.

Buyya, R., Yeo, C.S., Venugopal, S., Broberg, J., & Brandic, I. (2009). Cloud computing and emerging it platforms: Vision, hype, and reality for delivering computing as the 5th utility. *Future Generation Computer Systems, 25*(6), 599–616.

Casino, F., Dasaklis, T.K., & Patsakis, C. (2019). A systematic literature review of blockchain-based applications: Current status, classification and open issues. *Telematics and Informatics, 36*, 55–81.

Chen, H., Chiang, R.H., & Storey, V.C. (2012). Business intelligence and analytics: From big data to big impact. *MIS Quarterly*, 1165–1188.

Dolgui, A., & Ivanov, D. (2021). Ripple effect and supply chain disruption management: new trends and research directions. *International Journal of Production Research, 59*(1), 102–109.

Erkoyuncu, J.A., Butala, P., Roy, R., et al. (2018). Digital twins: Understanding the added value of integrated models for through-life engineering services. *Procedia Manufacturing, 16*, 139–146.

Hashem, I.A.T., Yaqoob, I., Anuar, N.B., Mokhtar, S., Gani, A., & Khan, S.U. (2015). The rise of "big data" on cloud computing: Review and open research issues. *Information Systems, 47*, 98–115.

Hulea, M., Rosu, O., Miron, R., & Aştilean, A. (2018). Pharmaceutical cold chain management: Platform based on a distributed ledger. In *2018 IEEE International Conference on Automation, Quality and Testing, Robotics (AQTR)* (pp. 1–6). IEEE.

Ivanov, D., & Dolgui, A. (2021). A digital supply chain twin for managing the disruption risks and resilience in the era of industry 4.0. *Production Planning & Control, 32*(9), 775–788.

Ivanov, D., Dolgui, A., Das, A., & Sokolov, B. (2019). Digital supply chain twins: Managing the ripple effect, resilience, and disruption risks by data-driven optimization, simulation, and visibility. *Handbook of Ripple Effects in the Supply Chain*, 309–332.

Ivanov, D., Dolgui, A., & Sokolov, B. (2022). Cloud supply chain: Integrating Industry 4.0 and digital platforms in the "supply chain-as-a-service". *Transportation Research Part E: Logistics and Transportation Review, 160*, 102676.

Javaid, M., Haleem, A., & Suman, R. (2023). Digital twin applications toward Industry 4.0: A review. *Cognitive Robotics, 3*, 71–92.

Jiang, Y., Yin, S., Li, K., Luo, H., & Kaynak, O. (2021). Industrial applications of digital twins. *Philosophical Transactions of the Royal Society A, 379*(2207), 20200360.

Kamble, S.S., Gunasekaran, A., & Sharma, R. (2020). Modeling the blockchain enabled traceability in agriculture supply chain. *International Journal of Information Management, 52*, 101967.

Kamble, S.S., Gunasekaran, A., Parekh, H., Mani, V., Belhadi, A., & Sharma, R. (2022). Digital twin for sustainable manufacturing supply chains: Current trends, future perspectives, and an implementation framework. *Technological Forecasting and Social Change, 176*, 121448.

Katsaliaki, K., Galetsi, P., & Kumar, S. (2022). Supply chain disruptions and resilience: a major review and future research agenda. *Annals of Operations Research*, 1–38

Kshetri, N. (2018). 1 blockchain's roles in meeting key supply chain management objectives. *International Journal of Information Management, 39*, 80–89.

Le, T. V., Li, X., & Le, T. A. (2024). *National digital twins for logistics and supply chain systems: Outlooks, components, enablers, and architecture*. SSRN (Social Science Research Network). http://dx.doi.org/10.2139/ssrn.5042679

Lepenioti, K., Bousdekis, A., Apostolou, D., & Mentzas, G. (2020). Prescriptive analytics: Literature review and research challenges. *International Journal of Information Management, 50*, 57–70.

Li, L., Aslam, S., Wileman, A., & Perinpanayagam, S. (2021). Digital twin in aerospace industry: A gentle introduction. *IEEE Access, 10*, 9543–9562.

Lim, K.Y.H., Chen, C.H., et al. (2024). Incorporating supply and production digital twins to mitigate demand disruptions in multi-echelon networks. *International Journal of Production Economics, 273*, 109258.

Macal, C., & North, M. (2014). Introductory tutorial: Agent-based modeling and simulation. In *Proceedings of the Winter Simulation Conference 2014* (pp. 6–20). IEEE.

Maheshwari, P., Kamble, S., Belhadi, A., Venkatesh, M., & Abedin, M.Z. (2023). Digital twin-driven real-time planning, monitoring, and controlling in food supply chains. *Technological Forecasting and Social Change, 195*, 122799.

Marston, S., Li, Z., Bandyopadhyay, S., Zhang, J., & Ghalsasi, A. (2011). Cloud computing—the business perspective. *Decision Support Systems, 51*(1), 176–189.

Mhaskey, S. V. (2024). Real-time risk management in supply chains using digital twins. *IUP Journal of Supply Chain Management, 21*(4), 29. http://dx.doi.org/10.71329/iupjscm/2024.21.4.29-42

Mihai, S., Yaqoob, M., Hung, D.V., Davis, W., Towakel, P., Raza, M., Karamanoglu, M., Barn, B., Shetve, D., Prasad, R.V., et al. (2022). Digital twins: A survey on enabling technologies, challenges, trends and future prospects. *IEEE Communications Surveys & Tutorials, 24*(4), 2255–2291.

Min, H. (2019). Blockchain technology for enhancing supply chain resilience. *Business Horizons, 62*(1), 35–45.

Minerva, R., Lee, G.M., & Crespi, N. (2020). Digital twin in the IoT context: A survey on technical features, scenarios, and architectural models. *Proceedings of the IEEE, 108*(10), 1785–1824.

Muhic, I., & Hodzic, M. (2014). Internet of things: Current technological review. *Periodicals of Engineering and Natural Sciences (PEN), 2*(2). Available online at: http://pen.ius.edu.ba

Nguyen, P., Ivanov, D., & Sgarbossa, F. (2023). A digital twin–based approach to reinforce supply chain resilience: simulation of semiconductor shortages. In *IFIP International Conference on Advances in Production Management Systems* (pp. 563–576). Springer.

Ogunsoto, O.V., Olivares-Aguila, J., & ElMaraghy, W. (2025). A conceptual digital twin framework for supply chain recovery and resilience. *Supply Chain Analytics, 9*, 100091.

Owusu-Berko, L. (2025). Advanced supply chain analytics: Leveraging digital twins, IoT and blockchain for resilient, data-driven business operations. *World Journal of Advanced Research and Reviews, 25*(02), 1777–1799. https://doi.org/10.30574/wjarr.2025.25.2.0572

Panigrahi, B.S., Vanitha, A., Palav, M.R., Babu, S.B.T., Nair, A.M., & Bogiri, N. (2024). IoT applications in cold chain management for pharmaceuticals: Ensuring product integrity and safety. In *2024 5th International Conference on Recent Trends in Computer Science and Technology (ICRTCST)* (pp. 66–70). IEEE.

Qi, Q., Tao, F., Hu, T., Anwer, N., Liu, A., Wei, Y., Wang, L., & Nee, A.Y. (2021). Enabling technologies and tools for digital twin. *Journal of Manufacturing Systems, 58*, 3–21.

Rehman, Z., Tariq, N., Moqurrab, S.A., Yoo, J., & Srivastava, G. (2024). Machine learning and internet of things applications in enterprise architectures: Solutions, challenges, and open issues. *Expert Systems, 41*(1), e13467.

Rinaldi, M., Caterino, M., Riemma, S., Macchiaroli, R., & Fera, M. (2025) Emergency supply chain resilience enhanced through blockchain and digital twin technology. *Logistics, 9*(1), 43.

Roca Zúñiga, A. (2023). Digital twins to assess inventory optimization. B.S. thesis, Universitat Politècnica de Catalunya.

Rojas, L., Peña, Á., & Garcia, J. (2025). AI-driven predictive maintenance in mining: A systematic literature review on fault detection, digital twins, and intelligent asset management. *Applied Sciences, 15*(6), 3337.

Saberi, S., Kouhizadeh, M., Sarkis, J., & Shen, L. (2019). Blockchain technology and its relationships to sustainable supply chain management. *International Journal of Production Research, 57*(7), 2117–2135.

Sandén, E., & Falk, J. (2021). *Digital twin-an exploratory study on its opportunities and challenges from a supply chain perspective*. Lund University, Department of Industrial Management and Logistics, Master's Thesis, Available online at: https://lup.lub.lu.se/student-papers/search/publication/9041669

Sarimveis, H., Patrinos, P., Tarantilis, C.D., & Kiranoudis, C.T. (2008). Dynamic modeling and control of supply chain systems: A review. *Computers & Operations Research, 35*(11), 3530–3561.

Sarker, D. (2023). Achieving digital transformation success: Interplay of performance, business models and digital twins in supply chain management. Master's thesis, NTNU.

Sharda, R., Delen, D., Turban, E., Aronson, J., & Liang, T. (2014). Business intelligence and analytics. *System for Decision Support, 398*, 2014.

Shashi, M. (2022). Digitalization of pharmaceutical cold chain systems using iotIoT digital enabler. *International Journal of Engineering and Advanced Technology (IJEAT), 11*(5), ISSN: 2249-8958 (Online).

Soykan, B., & Rabadi, G. (2024). Optimizing humanitarian logistics with deep reinforcement learning and digital twins. In *2024 Annual Modeling and Simulation Conference (ANNSIM)* (pp. 1–13). https://doi.org/10.23919/ANNSIM61499.2024.10732882

Soykan, B., Abdelnabi, A.B., & Rabadi, G. (2023). Adaptive scheduling in dynamic environments: A data-driven approach with digital twins and deep reinforcement learning. In *Proceedings of the 2023 IEEE 3rd International Conference on Digital Twins and Parallel Intelligence (DTPI)*, Orlando, FL, USA (pp. 1–6). https://doi.org/10.1109/DTPI59677.2023.10365465

Soykan, B., Blanc, G., & Rabadi, G. (2025). A proof-of-concept digital twin for real-time simulation: Leveraging a model-based systems engineering approach. *IEEE Access*, 1–1. https://doi.org/10.1109/ACCESS.2025.3557367
Stergiou, C., Psannis, K.E., Kim, B.G., Gupta, B. (2018). Secure integration of IoT and cloud computing. *Future Generation Computer Systems, 78*, 964–975.
Swan, M. (2015) *Blockchain: Blueprint for a new economy*. O'Reilly Media, Inc.
Tao, F., Xiao, B., Qi, Q., Cheng, J., & Ji, P. (2022). Digital twin modeling. *Journal of Manufacturing Systems, 64*, 372–389.
Tozanli, Ö., & Saénz, M.J. (2022). Unlocking the potential of digital twins in supply chains. *MIT Sloan Management Review, 63*(4), 1–4.
Venkatesh, L. (2025). 4 digital twin technologies. *Efficient Energy Utilization and Emission Reduction Strategies in Plant Operations*, 52.
Waller, M. A., & Fawcett, S. E. (2013). Data science, predictive analytics, and big data: a revolution that will transform supply chain design and management. *Journal of Business Logistics, 34*, 77–84. https://doi.org/10.1111/jbl.12010
Wang, S., Ouyang, L., Yuan, Y., Ni, X., Han, X., & Wang, F.Y. (2019). Blockchain-enabled smart contracts: architecture, applications, and future trends. *IEEE Transactions on Systems, Man, and Cybernetics: Systems, 49*(11), 2266–2277.
Wang, T., Feng, K., Ling, J., Liao, M., Yang, C., Neubeck, R., & Liu, Z. (2024). Pipeline condition monitoring towards digital twin system: A case study. *Journal of Manufacturing Systems, 73*, 256–274.
Wang, Y., Wang, X., & Liu, A. (2020). Digital twin-driven supply chain planning. *Procedia Cirp, 93*, 198–203.
Werner, A., Zimmermann, N., & Lentes, J. (2019). Approach for a holistic predictive maintenance strategy by incorporating a digital twin. *Procedia Manufacturing, 39*, 1743–1751
Wolniak, R., & Grebski, W. (2023). *The concept of diagnostic analytics*. Scientific Papers of Silesian University of Technology, Organization and Management Series no. 175, Available online at: https://bibliotekanauki.pl/articles/58909220.pdf
Wright, L., & Davidson, S. (2020). How to tell the difference between a model and a digital twin. *Advanced Modeling and Simulation in Engineering Sciences, 7*, 1–13.
Yigitbas, E., Karakaya, K., Jovanovikj, I., & Engels, G. (2021). Enhancing human-in-the-loop adaptive systems through digital twins and VR interfaces. In *2021 International Symposium on Software Engineering for Adaptive and Self-Managing Systems (SEAMS)* (pp. 30–40). IEEE.
Zaidi, S.A.H., Khan, S.A., & Chaabane, A. (2024). Unlocking the potential of digital twins in supply chains: A systematic review. *Supply Chain Analytics, 7*, 100075. https://doi.org/10.1016/j.sca.2024.100075
Zanella, A., Bui, N., Castellani, A., Vangelista, L., & Zorzi, M. (2014). Internet of things for smart cities. *IEEE Internet of Things Journal, 1*(1), 22–32.
Zhang, G., MacCarthy, B.L., & Ivanov, D. (2022). Chapter 5 - the cloud, platforms, and digital twins—enablers of the digital supply chain. In MacCarthy, B.L., Ivanov, D. (Eds.), *The digital supply chain* (pp. 77–91). Elsevier. https://doi.org/10.1016/B978-0-323-91614-1.00005-8

Chapter 6
Modeling Supply Chain Ecosystem Architecture to Support Digital Twins

Ola G. Batarseh

Abstract Supply chains are becoming increasingly complex and vulnerable to disruptions, from geopolitical conflicts to global health crises. In response, a model-based approach is proposed to architect and analyze supply chain ecosystems. This chapter explores how Model-Based Systems Engineering (MBSE) can be applied to supply chains through the Unified Architecture Framework (UAF). By leveraging UAF's enterprise viewpoints and aspects, supply chain architects can create a traceable, interconnected digital architectural model that enables early-stage validation, predictive analysis, and digital twin development.

The chapter introduces the enterprise architecture framework UAF, highlighting its viewpoint: strategic, operational, resources, security, and others, along with their cross-allocation. Using Dassault Systèmes' CATIA Magic Systems of Systems Architect (MSOSA), the chapter walks the reader through digital views of the supply chain, demonstrating how they support real-time inventory management, logistics coordination, cybersecurity controls, and roadmap planning. The ability to link enterprise elements through a digital thread provides a maintainable architecture along with actionable insights.

By replacing fragmented, document-based approaches with a single authoritative source of truth (ASOT), this model-based approach strengthens supply chain resilience, enhances adaptability, and enables organizations to systematically manage risks and disruptions. The chapter concludes by discussing scalability, reusability, and the role of enterprise architecture in supporting digital twin adoption for modern supply chains.

Keywords Model-Based Systems Engineering (MBSE) · Unified Architecture Framework (UAF) · Digital Twin · Supply Chain Architecture · Authoritative Source of Truth (ASOT)

O. G. Batarseh (✉)
Dassault Systemes, Chennai, Tamil Nadu, India

University of Massachusetts Lowell, Lowell, MA, USA

G. Rabadi, B. Soykan (eds.), *Optimizing Supply Chains Through Digital Twins*, International Series in Operations Research & Management Science 375,
https://doi.org/10.1007/978-3-032-08147-6_6

Supply chains are becoming increasingly complex and vulnerable to disruptions, from geopolitical conflicts to global health crises. In response, a model-based approach is proposed to architect and analyze supply chain ecosystems. This chapter explores how Model-Based Systems Engineering (MBSE) can be applied to supply chains through the Unified Architecture Framework (UAF). By leveraging UAF's enterprise viewpoints and aspects, supply chain architects can create a traceable, interconnected digital architectural model that enables early-stage validation, predictive analysis, and digital twin development.

The chapter introduces the enterprise architecture framework UAF, highlighting its viewpoint: strategic, operational, resources, security, and others, along with their cross-allocation. Using Dassault Systèmes' CATIA Magic Systems of Systems Architect (MSOSA), the chapter walks the reader through digital views of the supply chain, demonstrating how they support real-time inventory management, logistics coordination, cybersecurity controls, and roadmap planning. The ability to link enterprise elements through a digital thread provides a maintainable architecture along with actionable insights.

By replacing fragmented, document-based approaches with a single authoritative source of truth (ASOT), this model-based approach strengthens supply chain resilience, enhances adaptability, and enables organizations to systematically manage risks and disruptions. The chapter concludes by discussing scalability, reusability, and the role of enterprise architecture in supporting digital twin adoption for modern supply chains.

6.1 Modeling Supply Chain Ecosystem Architecture to Support Digital Twins

Supply chains are becoming more complex and increasingly vulnerable to disruptions, whether due to global health crises or regional conflicts. In this era of digital twins, adopting a model-based approach to digitize supply chain ecosystem architecture is not only timely but essential. Supply chain practitioners and researchers stand to gain significantly from understanding the model-based practices advanced by systems engineers in managing complex systems, such as aircraft design. These approaches, referred to as Model-based System Engineering (MBSE), can offer powerful insights for building resilient, adaptive supply chain systems.

MBSE has significantly transformed the aircraft industry by providing an authoritative source of truth (ASOT) for system architecture, replacing the fragmented document-based approach. By using a standardized and unified language, MBSE facilitates clear communication among architects, engineers, and computer systems. When the ASOT can be analyzed by computers, it unlocks access to advanced tools like digital twins, artificial intelligence (AI), and more.

In addition, referencing the ASOT across the different teams and disciplines has improved collaborations in designing complex systems, enabling deeper insights, predictive analytics, and smarter decision-making. The model-based practices have provided a virtual environment enabling early-stage validation and verification, allowing for the simulation and analysis of system performance and early error detection. The ability to validate requirements before physical prototyping helps reduce the number of design iterations, shortening development timelines and lowering costs, all while complying with regulatory and safety standards.

The aircraft industry has been employing MBSE practices for over 2 decades. This chapter aims to shed light on these practices and explore how they can be applied to supply chains using the Unified Architecture Framework (UAF) by Object Management Group (OMG). The chapter begins with introducing the enterprise architecture of UAF. Based on these principles, the chapter discusses how UAF can support a structured model-based supply chain model. The chapter concludes with how these models support decision-making and enable digital twin adoption.

The Unified Architecture Framework (UAF) is one of the most modern enterprise architecture frameworks developed by the Object Management Group (OMG) (Object Management Group, 2022). An enterprise architecture consists of a network of interdependent resources—such as people, processes, organizations, supporting technologies, and funding—that:

- interact with each other to coordinate functions,
- share information,
- allocate funding,
- create workflows, and
- make decisions.

to achieve business and operational goals through a complex web of interactions distributed across geography and time (Rebovich & White, 2011). Unlike a single solution architecture, UAF encompasses the needs and functions of diverse user communities, including acquisition organizations, end users, contractors, and the engineering and program offices.

The Unified Architecture Framework (UAF) is based on the Unified Modeling Language (UML) (Object Management Group, 2017) and the Systems Modeling Language (SysML) (Object Management Group, 2019), making it a model-based modeling language. UAF's modernization in 2017 builds upon legacy enterprise architecture frameworks that have been widely adopted by defense organizations, such as the Ministry of Defence Architecture Framework (MODAF) (Ministry of Defence, 2014), NATO Architecture Framework (NAF) (NATO Communications and Information Agency, 2017), and Department of Defense Architecture Framework (DoDAF) (Department of Defense, 2010). UAF has been tailored to serve both defense and commercial applications, ensuring versatility across various industries. Figure 6.1 illustrates the maturity levels that have evolved throughout the history of these frameworks, presented using the Project Roadmap view in UAF. This figure demonstrates how UAF stands as the most modern enterprise architecture framework by OMG, offering the most comprehensive set of viewpoints compared to the historical frameworks from which it evolved.

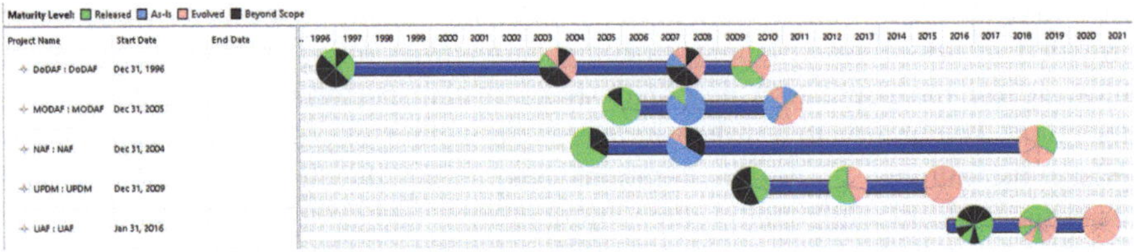

Fig. 6.1 Evolution of enterprise architecture framework maturity levels, visualized using the UAF Project Roadmap view

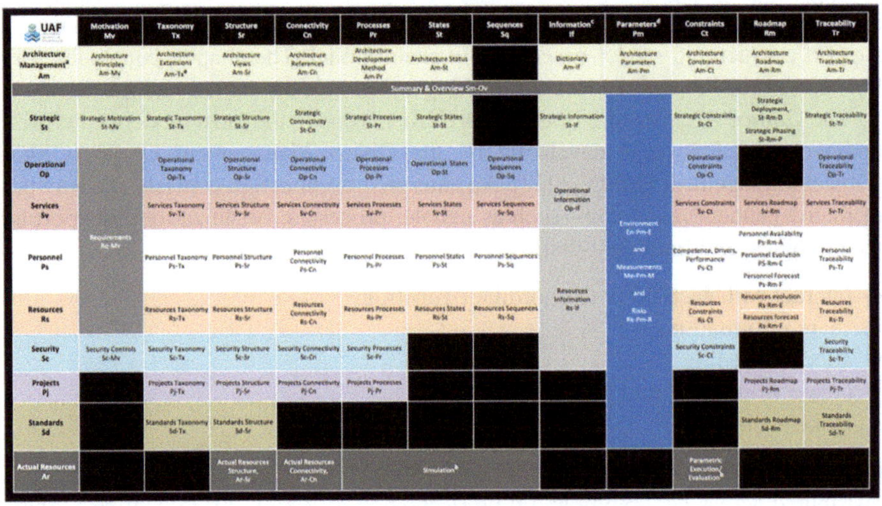

Fig. 6.2 UAF grid showing how the various view specifications (cells) correspond to viewpoints (prev. Known as domains) (horizontal rows) and the aspects (prev. Known as model kinds) (the columns)

Due to the complexity of managing the multiple viewpoints with overlapping concerns and metamodels, the standard viewpoints are refactored into a more manageable format shown in the grid as in Fig. 6.2 (Object Management Group, 2022).

The UAF aspects provide a comprehensive way for architects to address different layers of complexity in a structured, modular manner. These aspects—Motivation, Taxonomy, Structure, Connectivity, Behavior, Information, Constraints, Roadmaps, Analytics, and Traceability—enable a holistic view across various viewpoints, enhancing decision-making and strategic alignment. For further details on modeling views that reflect these aspects across the viewpoints, refer to the UAF metamodel outlined in the references section (Object Management Group, 2022).

Perhaps the most compelling aspect of UAF is the range of viewpoints it offers enterprise architects. Within a single digital environment, architects can create interconnected digital views that capture, trace, and relate the following enterprise perspectives (Object Management Group, 2022):

1. Strategic Viewpoint: Models the overarching strategic goals, objectives, and capabilities required to fulfill the mission of the enterprise.
2. Operational Viewpoint: Defines the operational performer, activities, and exchanges necessary to achieve the strategic goals.
3. Services Viewpoint: Describes the services that support operational processes, detailing service interactions and relationships.
4. Personnel Viewpoint: Models the human resources involved in the enterprise, responsibilities, and competencies.
5. Resources Viewpoint: Focuses on physical and software resources that support operations.
6. Security Viewpoint: Captures security controls and processes to protect assets, information, and systems within the enterprise.
7. Projects Viewpoint: Defines the lifecycle of projects, project dependencies, and milestones associated with exhibiting capabilities.
8. Standards Viewpoint: Identifies standards, policies, and regulations that govern enterprise operations and ensure compliance.
9. Actual Resources Viewpoint: Provides a detailed view of the current state of resources, including their allocation and utilization within the enterprise. This viewpoint also enables parametric analysis, allowing architects to perform comprehensive enterprise architecture evaluations.

This rich metamodel of the enterprise architecture allows architects to model a complete architecture ecosystem. Let's take a supply chain architecture as an example and walk through the implementation of these viewpoints, illustrated through snapshots of digital views created in Dassault Systèmes' CATIA Magic Systems of Systems Architect (MSOSA) software (Dassault Systèmes, 2021).

The electronics supply chain is one of the most complex and interdependent systems globally. Its complexity arises from the geographic spread of suppliers, manufacturers, logistics providers, and customers across continents. In addition, it relies on intricate processes to deliver high-tech products like smartphones and laptops (Sweeney & Mitchell, 2020).

The following digital views in UAF have been created to represent the enterprise viewpoints of a fictional electronics supply chain. These views were developed to showcase how UAF can be applied to model the complexities and interdependencies within such a supply chain, which in reality includes suppliers, manufacturers, logistics providers, and customers across various geographical regions. It is important to emphasize that these views, within a digital modeling environment, are traceable and executable, enabling detailed analysis and simulation. Unlike static tools like Visio, which also employ standard languages for communication, UAF digital models go further by providing actionable insights and dynamic interconnections, making them indispensable for architects aiming to address the complexities of modern supply chains effectively. Figure 6.3 illustrates the connection between the organization's vision, the design phase of the supply chain system, the capabilities

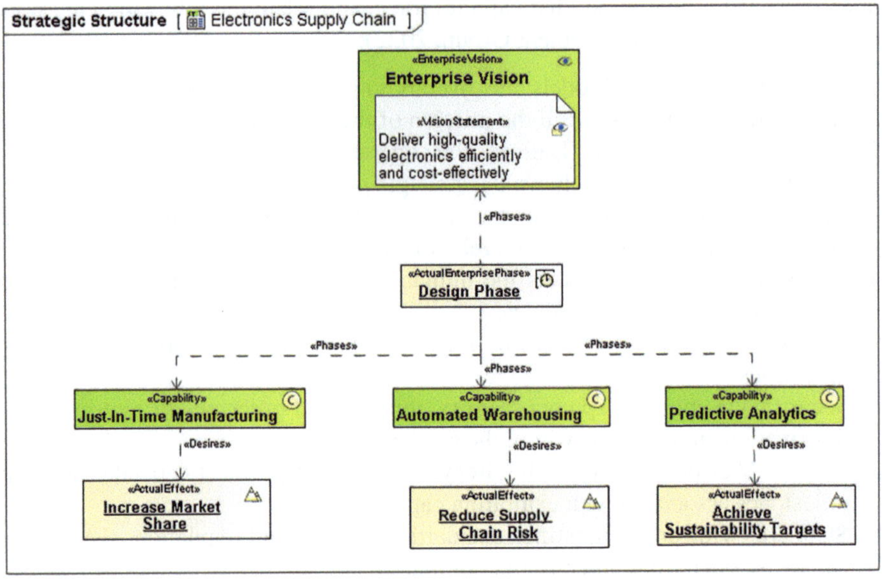

Fig. 6.3 Vision, design phase elements, capabilities, and actual effects alignment

required to achieve strategic goals, and the actual effects realized through their implementation. It highlights the integration of strategic planning with actionable design elements to ensure alignment between goals and outcomes. This figure demonstrates a limited subset of model elements compared to the full range that could be captured, such as value streams, desired effects, value items, challenges, and opportunities. Incorporating these elements would provide a more comprehensive representation of the strategic viewpoint and its benefits.

The realization of these capabilities necessitates the development of a Concept of Operations (ConOps) by identifying operational performers. These performers will later be traced to the resources that exhibit the required capabilities. To begin, we construct the Operational Connectivity Diagram, which provides a high-level view of interactions between performers through item flows. Figure 6.4 helps establish the foundation for mapping how operational interactions support strategic goals and capabilities, ensuring alignment across different layers of the architecture.

This operational connectivity diagram differs from traditional document-based platforms like Visio or PowerPoint by being digitally enabled within a model-based environment. It starts with ConOps information flows and links each supplier interaction to the processes producing and consuming these flows. Each process is tied to detailed activity diagrams outlining resources and steps, enabling simulation within the same digital environment. This traceability supports architecture optimization and facilitates the creation of a supply chain digital twin, enhancing decision-making and operational efficiency.

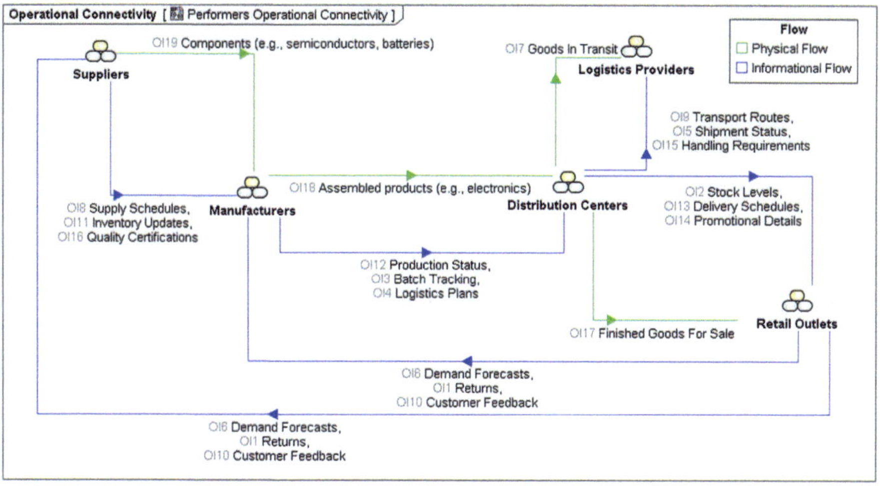

Fig. 6.4 High-level item flows between supply chain performers

#	Exchan... ID	Operational Exchange Item	Sending Operational Performer	Receiving Operational Performer	Producing Operational Activity	Consuming Operational Activity
1	OE1	OI19 Components (e.g., semiconductors, batteries)	Suppliers	Manufacturers	Component Sourcing and Manufacturing	Assembly and Production Planning
2	OE2	OI8 Supply Schedules OI11 Inventory Updates OI16 Quality Certifications	Suppliers	Manufacturers	Supplier Planning and Quality Assurance	Production Scheduling and Inventory Management
3	OE3	OI18 Assembled products (e.g., electronics)	Manufacturers	Distribution Centers	Final Assembly and Packaging	Warehouse Management and Distribution
4	OE4	OI12 Production Status OI3 Batch Tracking OI4 Logistics Plans	Manufacturers	Distribution Centers	Production Monitoring and Batch Tracking	Logistics Coordination and Inventory Distribution
5	OE5	OI17 Finished Goods For Sale	Distribution Centers	Retail Outlets	Warehouse Operations and Order Fulfillment	Retail Operations and Marketing Planning
6	OE6	OI2 Stock Levels OI13 Delivery Schedules OI14 Promotional Details	Distribution Centers	Retail Outlets	Inventory Management and Delivery Coordination	Retail Operations and Sales Channels
7	OE7	OI7 Goods In Transit	Distribution Centers	Logistics Providers	Shipment Preparation and Dispatch	Logistics Execution and Transport Management
8	OE8	OI9 Transport Routes OI5 Shipment Status OI15 Handling Requirements	Distribution Centers	Logistics Providers	Shipment Coordination and Route Planning	Transport Management and Compliance
9	OE9	OI6 Demand Forecasts OI1 Returns OI10 Customer Feedback	Retail Outlets	Manufacturers	Market Demand Analysis and Customer Service	Production Planning and Supply Chain Optimization
10	OE10	OI6 Demand Forecasts OI1 Returns OI10 Customer Feedback	Retail Outlets	Suppliers	Market Demand Analysis and Customer Service	Production Planning and Supply Chain Optimization

Fig. 6.5 Mapping information flows to producing and consuming processes

Let's take a look at Fig. 6.5, which illustrates each information flow in the operational connectivity diagram. This figure maps the specific processes that produce and consume these information flows between operational performers, such as suppliers, manufacturers, and logistics providers. It highlights the dynamic interaction network, enabling traceability to detailed activity diagrams for deeper insights into the supply chain operations and how they integrate to drive overall performance.

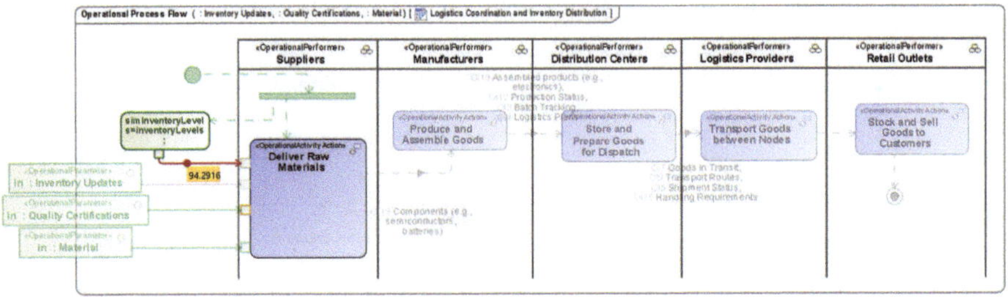

Fig. 6.6 Logistics coordination and inventory distribution process flow sample

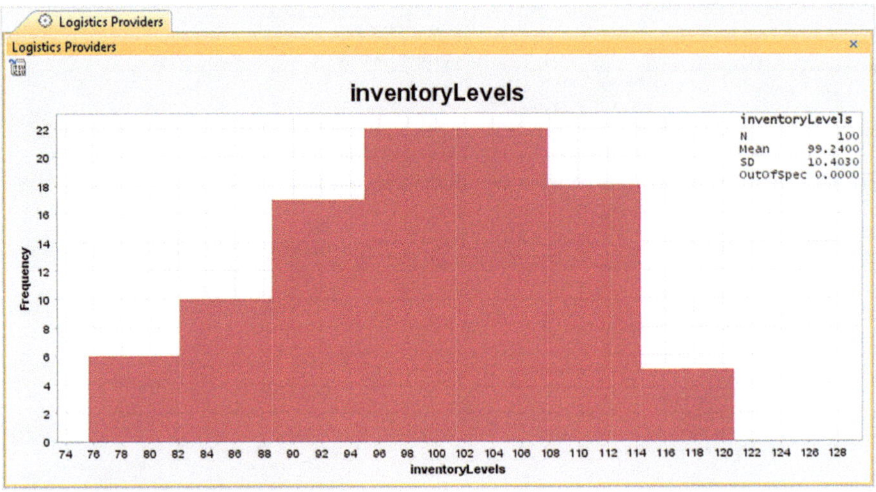

Fig. 6.7 Inventory levels histogram

We can explore modeling the logistics planning process directly within the digital environment. Figure 6.6 illustrates a streamlined interaction process among operational performers for logistics coordination and inventory distribution during execution. The focus is not on detailed logic modeling, which the tool supports, but on showcasing the tool's architecting capabilities. In Fig. 6.6, the information exchanges are displayed between the activities that produce and consume them, keeping the model viewpoints digitally connected. Figure 6.7 features a histogram of inventory levels from the activity simulation, demonstrating how embedded analytics influence decision-making early in the design phase.

Behavioral modeling in UAF supports state and activity synchronization, linking tasks such as demand forecasting, route planning, inventory preparation, and logistics coordination. Figure 6.8 illustrates the state machine of Logistics Providers, detailing its states and transitions triggered by signals from other supply chain performers. This model also showcases how Logistics Providers control and influence related activities. By visualizing state triggers and behavior effects,

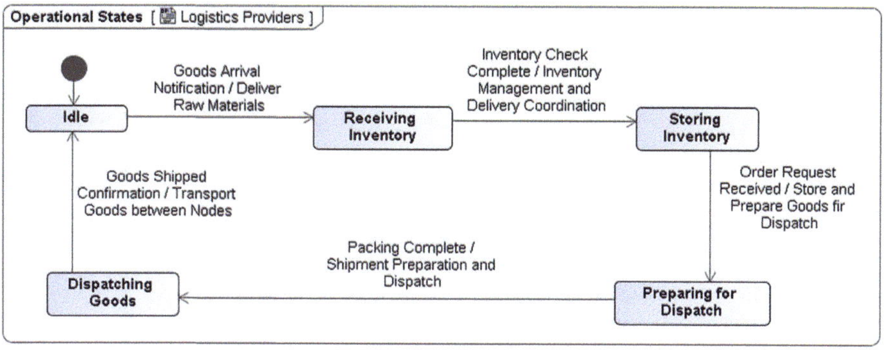

Fig. 6.8 State machine of logistics providers

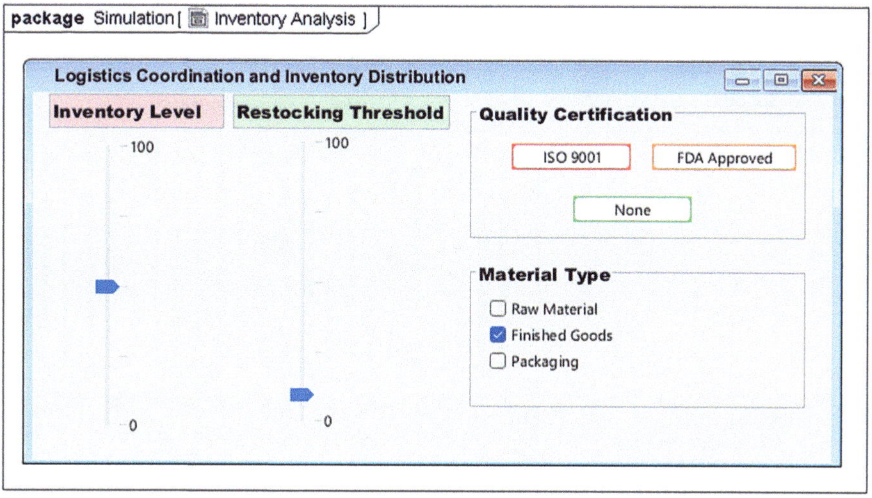

Fig. 6.9 Simulation control panel for inventory management in CATIA Magic

this synchronized view supports early architecture decisions, helping avoid costly changes later in the lifecycle.

In addition, the simulation in this architectural tool can be controlled through GUIs, enabling early-stage design analysis with executable scenarios. Identifying issues early through simulation helps reduce the risk of expensive corrections later in the design process. Figure 6.9 illustrates a GUI example that allows users to select various options, each influencing different analyses and ultimately impacting decision-making.

Besides early simulation capabilities, these model-based systems engineering tools enable object-oriented modeling, offering powerful capabilities to create visual diagrams using standard languages. These diagrams gain greater value when traced and mapped to enterprise elements across domains. For example, the traceability

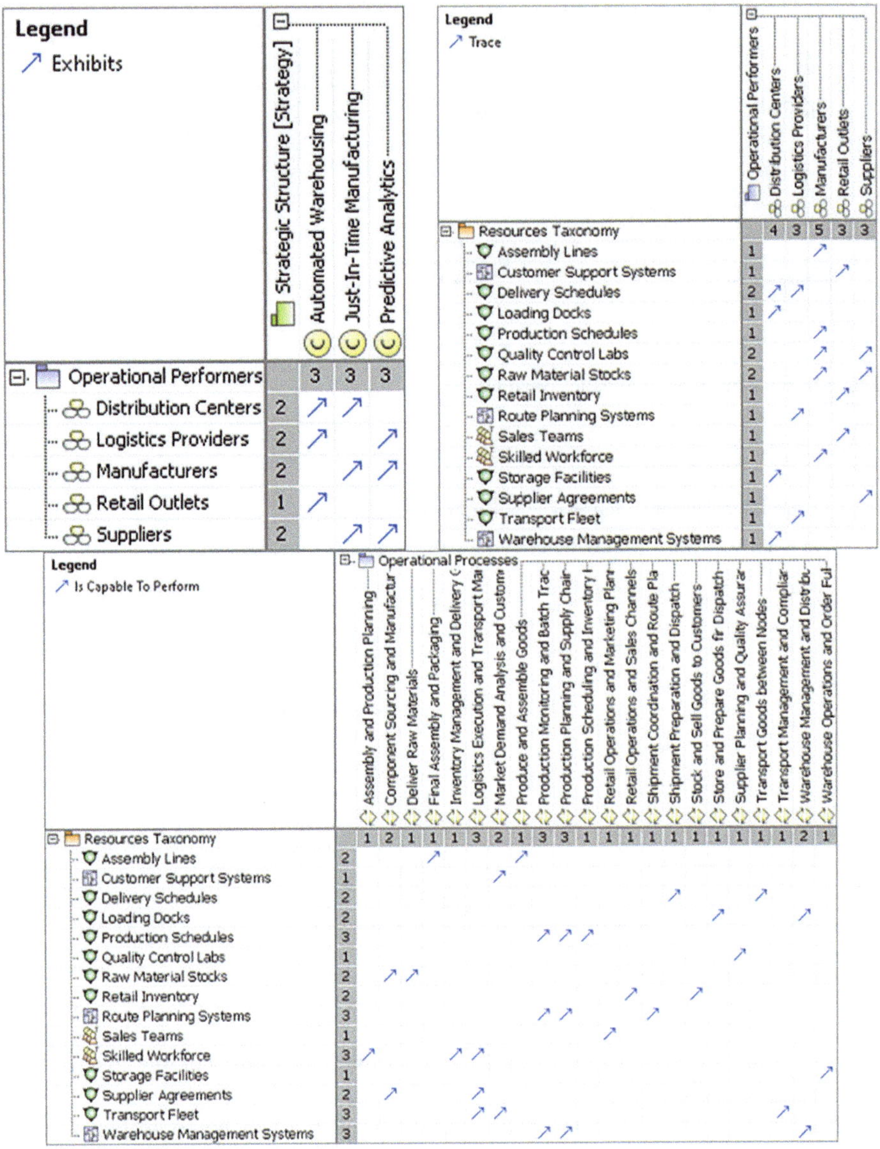

Fig. 6.10 Traceability matrices across enterprise elements in different domains

matrices, in Fig. 6.10, link operational performers modeled in the ConOps to the capabilities they exhibit and the resources supporting operational activities. Cross-allocating enterprise elements within a digital environment like CATIA Magic enhances strategic and operational decision-making early in the design phase,

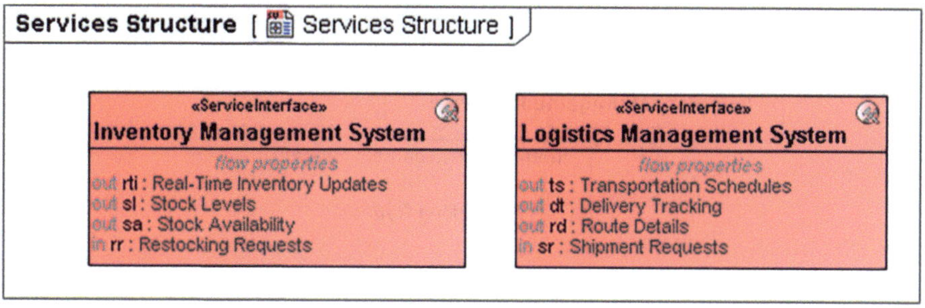

Fig. 6.11 Service interfaces definition

while digital traceability maintains a continuous thread across enterprise elements, ensuring seamless integration and alignment throughout the system's lifecycle.

This chapter does not delve into the specifics of enterprise architecture modeling using UAF, but rather illustrates the capabilities of modeling these semantics in a digital format. In addition to the strategic, operational, and resource viewpoints discussed earlier, the framework also offers the capability to model the personnel components of the supply chain. For example, it enables the creation of organizational charts that include actual people, roles, competencies, responsibilities, and their connection to the capabilities within the strategic domains.

In addition, the framework allows modelers to define both provided and required services within the enterprise, illustrating how these services support operational activities and strategic capabilities. The power of interface modeling is fully leveraged in the services domain, enabling the mapping of item flows across interfaces. Interface reuse or modularity enhances efficiency and ensures consistency across various systems and interactions. This approach facilitates flexibility, scalability, and seamless integration between different enterprise components, especially in modern, interconnected supply chains. Figure 6.11 displays the service interfaces along with their flow properties, including their direction and type. Figure 6.12 illustrates how these interfaces are connected within the architecture, highlighting the flow elements and their relationships.

The Security Domain in UAF supports the modeling of critical elements to protect the enterprise, namely, security controls, risk mitigation strategies, and protective measures for assets. This viewpoint also allows for defining and allocating responsibilities to organizational resources tasked with executing these controls. Specifically for supply chains, it addresses vulnerabilities such as cyber risks, physical threats, and operational disruptions. In today's era of cybersecurity, modeling these aspects within enterprise architecture is essential to ensure resilience, compliance, and the ability to adapt to evolving threats systematically. Figure 6.13 shows two elements from the security domain: security controls and their role in protecting enterprise services. It also illustrates the associated security risks and identifies the organizations responsible for managing these risks.

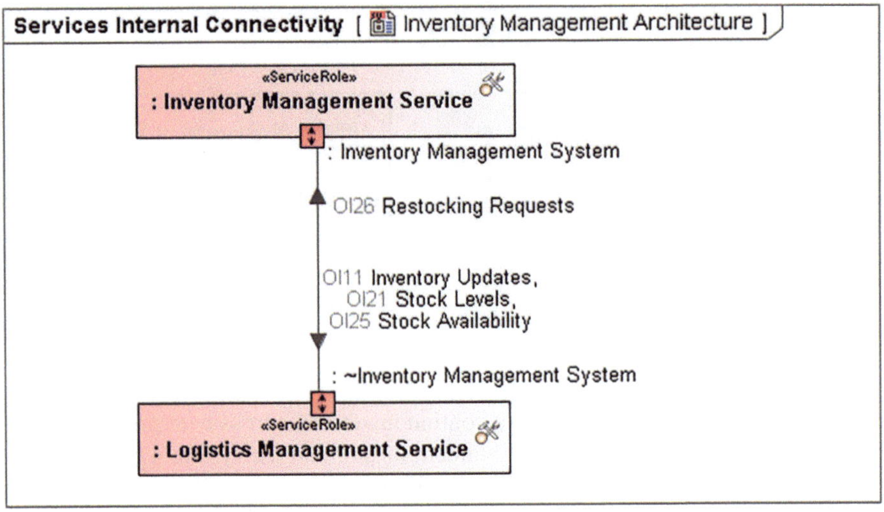

Fig. 6.12 Usage of interfaces as ports displaying item flows

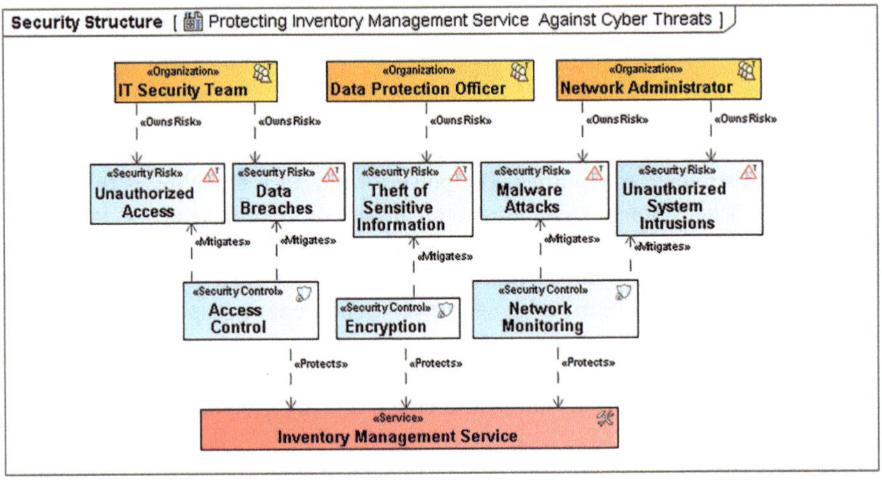

Fig. 6.13 Implementing security viewpoints with cross-allocation

Figure 6.13 provides an excellent example of cross-allocation between UAF viewpoints. It highlights elements from personnel, security, and services viewpoints, showcasing their interrelationships. This cross-allocation is crucial for traceability analysis across viewpoints. For instance, a change in the service viewpoint may necessitate corresponding updates in the security domain or other related areas. Model-based approaches facilitate such digital traceability analysis across enterprise viewpoints, providing a cohesive alternative to fragmented, document-based methods.

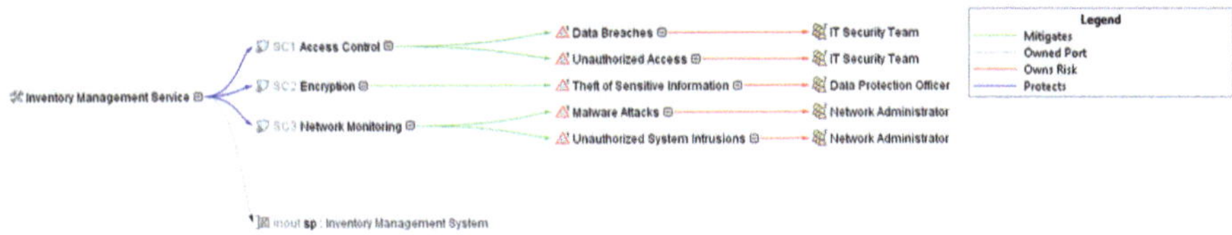

Fig. 6.14 Digital thread relationship map linking services, interfaces, security controls, and risk ownership

Generic tables, dependency matrices, and relationship maps are examples of general analysis views in MSOSA. These views enable the operationalization of architecture models. Additionally, the meta-chain analysis, which is readily available to users, facilitates decision-making early in the enterprise lifecycle design phase. Figure 6.5 illustrates an example of predefined tables in CATIA Magic that support architecture model analysis. These tables are customizable, enabling meta-chain analysis across the complete digital thread of the architecture. This, in turn, supports the development of the digital twin as a by-product of the architecture. By establishing a well-structured architecture model, the process of building a digital twin becomes more transparent and comprehensible, ensuring alignment between the physical and digital representations.

This digital thread from Services to Security Controls is structured across the following architectural layers:

- Service: Real-Time Inventory Management System (IMS).
- Interfaces: Inventory Update API (provided/required ports).
- Security Controls: Access Control, Encryption.
- Risk Owner: IT Security Team.
- Risk: Unauthorized access to inventory data.

Architects can extract and analyze these connected elements using relationship maps or structured tables for architecture verification, validation, and decision-making. Figure 6.14 illustrates a digital thread relationship map, which can also be represented in a tabular format within the model. Consistent modeling across domains ensures that these extracted views remain available for analysis and traceability.

In addition to the architectural views explained above, extracted from the enterprise viewpoints based on the UAF, namely, strategic, operational, resources, services, and others, the framework also supports modeling the evolution of architecture. This is achieved through model-based roadmaps, which are aligned with future capability acquisition efforts. The past, present, and future of the enterprise architecture are captured within the same model. The inclusion of time tags on architectural model elements further empowers MBSE tools, replacing static PowerPoint timeline slides with dynamic depictions of the intended architecture evolution.

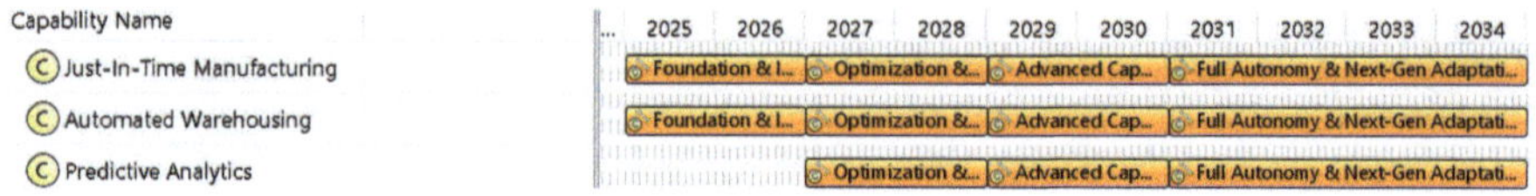

Fig. 6.15 Employing the digital roadmap strategic phasing from UAF

UAF incorporates roadmaps as one of its aspects, as shown in the 11th column of the UAF grid presented in Fig. 6.2. Different roadmaps are associated with the various viewpoints, which are represented in the rows of the UAF grid. For example, in the strategic domain, there are four available roadmap diagrams, including the strategic phasing roadmap discussed here and how it is connected to the project roadmap. Additionally, there are other roadmaps tailored to different enterprise aspects, such as personnel evolution, personnel availability, standards forecasting, and more. These architectural evolution views have been quickly adopted in various applications for the capability they provide in high-level executive decision-making, thanks to their digital traceability across all enterprise aspects.

For our example, the capability phasing is outlined using a strategic phasing roadmap displayed on a timeline similar to a Gantt chart, as shown in Fig. 6.15. The capabilities are listed on the left side of the roadmap, while the planned capability configurations for each planning year are represented as horizontal lines aligned with specific time points. Although it resembles a Gantt chart, this model-based and analytical roadmap is far more powerful, as it is traceable to various enterprise elements.

These horizontal bars represent capability configurations, each with its own architectural views detailing the resources that compose them and the interactions between these resources. Additionally, the release and withdrawal time points for these configurations are managed by architectural model elements known as actual project milestones. By connecting these milestones to the responsible resources, the architecture model ensures consistent evolution over time.

The milestones that are assigned to release and withdraw the capability configurations are also to be architected with status indicators that express their maturity. Figure 6.16 shows a snapshot of the milestone with an actual date, architected with four themes to assign maturity at different points in time:

- Inventory Turnover Ratio.
- Stockout Rate.
- Resource Utilization Rate.
- Cost Efficiency.

This milestone date enables the tracking and tracing of architectural evolution, ensuring traceability to the responsible resource and the capability configuration designed to exhibit the relevant capabilities. By maintaining the architecture of different enterprise elements, their instantiations, and their evolution in one integrated model, architects have access to readily available digital threads for informed

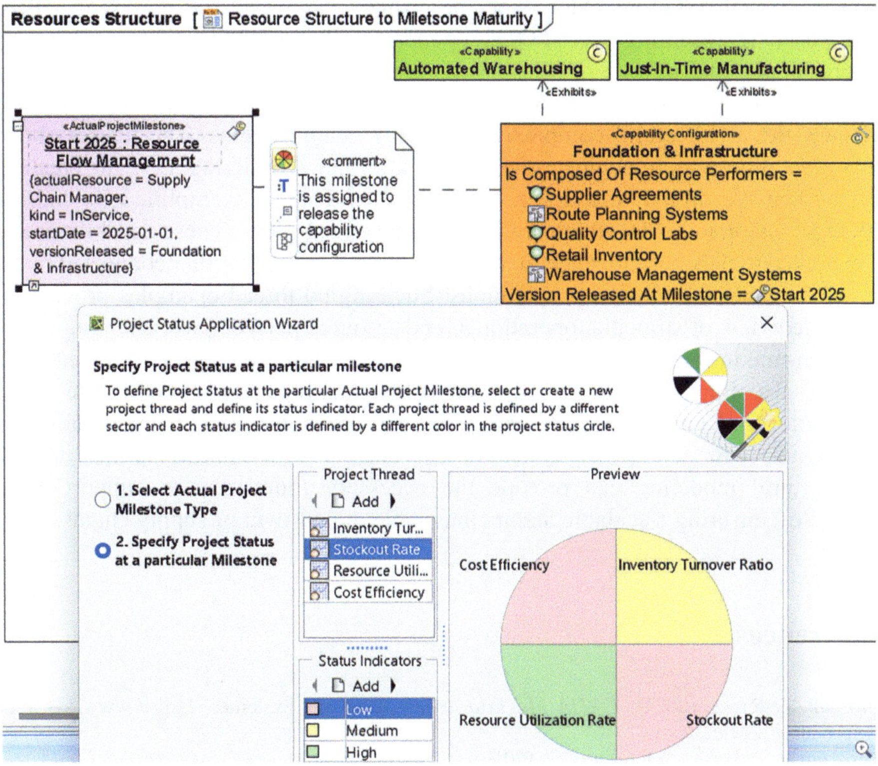

Fig. 6.16 Architecting supply chain milestones and their maturity in UAF

decision-making. This consistent approach eliminates the fragmentation typically caused by separate documents capturing different aspects of the enterprise.

This approach maintains the model as an ASoT, serving as a living artifact that seamlessly connects all enterprise elements and their evolution. It provides end-to-end traceability and adaptability, enabling dynamic updates to the digital thread whenever changes occur.

We present architectural modeling, guided by MBSE principles, as a critical enabler of digital twin development in supply chains. UAF architectural modeling provides a structured approach to defining enterprise elements, their interactions, and lifecycle dependencies and evolution, which are essential for developing a cohesive digital twin reliant on digital architecture.

In the 33rd Annual INCOSE International Symposium, Pierce et al. illustrated that "a detailed digital twin system model was created and maintained late in the development lifecycle, providing design insight in support of flight operations," an achievement not possible through traditional document-based approaches. Their approach validates the value of architectural modeling by demonstrating how an integrated ASoT, as presented by an MBSE model, reduces the time needed to answer critical questions. Moreover, the project showcased the power of MBSE and reuse, "representing over 9000 components with only 600 definition elements" (Pierce et al., 2023).

This chapter leverages the power of UAF as an MBSE framework to efficiently manage the architectural complexity of supply chain systems. The figures above represent digital architecture rather than static views, utilizing UAF to create a comprehensive digital thread for supply chains. This is exemplified by linking strategic vision to goals, which are connected to capabilities. These capabilities are then linked to the operational elements that exhibit them and further connected to the resources that implement the operations. Such digital thread examples provide an integrated view of strategic, operational, and resource perspectives, enabling better decision-making across the supply chain while also enhancing lifecycle traceability.

It is also important to note that an MBSE approach is based on the reuse of architectural elements while enhancing the scalability and maintainability of the digital twin model. We conclude this chapter by emphasizing that UAF's architectural modeling can provide the necessary foundation to support these principles, ensuring a scalable and maintainable digital twin of supply chains.

References

Dassault Systèmes. (2021). *CATIA Magic overview*. Dassault Systèmes. https://www.3ds.com/products-services/catia/

Department of Defense. (2010). *Department of Defense architecture framework (DoDAF) version 2.02*. Department of Defense https://www.af.mil/Portals/1/documents/DoDAF/DoDAF_v2.02.pdf

Martin, Bradley, Laura H. Baldwin, Paul DeLuca, Natalia Henriquez Sanchez, Mark Hvizda, Colin D. Smith, & N. Peter Whitehead (2023). *Supply chain interdependence and geopolitical vulnerability: The case of Taiwan and high-end semiconductors*. RAND Corporation. Available at: RAND Corporation.

Ministry of Defence. (2014). *MODAF (Ministry of Defence Architecture Framework)*. Ministry of Defence. https://www.gov.uk/government/publications/modaf

NATO Communications and Information Agency. (2017). *NATO Architecture Framework (NAF) v4* (Version 4.0). NATO Communications and Information Agency. https://www.nato.int/cps/en/natolive/topics_82795.htm

Object Management Group. (2017). *Unified modeling language (UML) specification* (Version 2.5.1). Object Management Group. https://www.omg.org/spec/UML/2.5.1/

Object Management Group. (2019). *Systems modeling language (SysML) specification* (Version 1.6). Object Management Group https://www.omg.org/spec/SysML/1.6/

Object Management Group. (2022). *Unified Architecture Framework (UAF) specification* (Version 1.2). Object Management Group https://www.omg.org/spec/UAF/1.2/

Pierce, G. J., Heeren, J. D., & Hill, T. R. (2023). Orion SysML Model, Digital Twin, and Lessons Learned for Artemis I. *INCOSE International Symposium, 33*(1), 290–304.

Rebovich, G., & White, B. E. (2011). *Enterprise systems engineering: Advances in theory and practice*. CRC Press, Francis Taylor Group, Auerbach.

Sweeney, D., & Mitchell, K. (2020). Understanding the complexities of the global electronics supply chain. *International Journal of Supply Chain Management, 15*(3), 34–47.

Chapter 7
A Supply Chain Digital Twin Architecture for Semiconductor Industry

Bulent Soykan, Sean Mondesire, Ghaith Rabadi, and Grace Bochenek

Abstract The semiconductor industry operates within a uniquely challenging environment characterized by extreme complexity, rapid technological evolution, and stringent quality requirements, rendering its global supply chains vulnerable to disruption. This chapter explores the application of digital twins (DTs) as a strategic solution to enhance operational management and resilience in this critical sector. We define the semiconductor supply chain DT as a dynamic virtual replica of the physical network, continuously synchronized through real-time data integration. The chapter details the core architecture, focusing on the central data hub and the integration of key components: foundational business systems (ERP, MES, PLM, CRM), specialized planning and dispatching services, advanced analytics and AI/ML tools, and vital external data streams from suppliers, customers, and IoT sensors. Key functionalities enabled by this integrated framework—including real-time monitoring, predictive simulation, and process optimization—are examined. We highlight the significant benefits derived, such as enhanced end-to-end visibility, proactive risk mitigation, improved operational agility, optimized resource allocation, and continuous quality assurance.

Keywords Digital Twin · Semiconductor industry · Supply chain management · System architecture · Data integration · Predictive analytics

B. Soykan (✉)
Institute for Simulation and Training, University of Central Florida, Orlando, FL, USA
e-mail: Bulent.Soykan@ucf.edu

S. Mondesire · G. Rabadi · G. Bochenek
School of Modeling, Simulation and Training, University of Central Florida, Orlando, FL, USA
e-mail: Sean.Mondesire@ucf.edu; Ghaith.Rabadi@ucf.edu; Grace.Bochenek@ucf.edu

G. Rabadi, B. Soykan (eds.), *Optimizing Supply Chains Through Digital Twins*, International Series in Operations Research & Management Science 375,
https://doi.org/10.1007/978-3-032-08147-6_7

7.1 Introduction

Global supply chains are essential to the functioning of modern economies, facilitating the optimized distribution of goods and services spanning borders (Gereffi & Lee, 2012). However, these interconnected systems face increasing vulnerability to frequent disruptions, ranging from natural disasters and geopolitical tensions to pandemics and logistical bottlenecks (Lund et al., 2020). Such disruptions carry significant consequences, including financial losses, threats to economic stability, and impacts on national security. Its intricate network, for instance, manages the global flow of critical inputs ranging from ultrapure silicon wafers and numerous specialty chemicals to essential process gases like neon, demanding meticulous control. The exact specifications and sensitive nature of these globally sourced materials significantly amplify the potential impact of any supply chain disturbance. The semiconductor industry, forming the bedrock of modern technology, particularly exemplifies these challenges due to its unique operating conditions.

This industry is marked by profound complexity (Aelker et al., 2013; Hicks, 1996). Its global supply chains involve numerous specialized suppliers, intricate manufacturing processes with hundreds of high-precision steps, and the complex coordination of vast data streams (Banerjee, 2007). Managing these elements effectively is a significant operational challenge, making the sector highly sensitive to any interruption. Also, the industry experiences rapid evolution, driven by relentless technological advancements and the demand for increasingly powerful chips (Karrila, 2022). This necessitates constant adaptation in manufacturing and supply chain agility, adding another layer of operational difficulty. In addition, semiconductor manufacturing adheres to extremely stringent quality demands. Minute imperfections can lead to critical failures, requiring meticulous process control, ultraclean environments, and rigorous testing (Mönch et al., 2018).

Given these inherent complexities, rapid changes, and critical quality requirements, disruptions within the semiconductor supply chain can have particularly severe cascading effects. In response to these vulnerabilities, industries and governments are increasingly exploring resilient solutions. One of the most promising approaches is the adoption of digital twins (DTs). DTs are data-driven virtual models designed to replicate physical supply chain components and processes in near real time (Friederich et al., 2022; Ivanov et al., 2019; Soykan et al., 2025). By integrating data from sensors, core business systems like enterprise resource planning (ERP) and manufacturing execution system (MES), and external sources, these dynamic models allow organizations to monitor operations comprehensively, predict potential disruptions before they occur, and simulate and optimize responses, thereby offering a powerful tool to navigate the demanding landscape of semiconductor manufacturing and logistics.

One of the key advantages of DTs is real-time monitoring (Qamsane et al., 2019). By leveraging advanced data analytics, machine learning (ML), and artificial intelligence (AI), DTs provide organizations with enhanced visibility into their supply networks. This visibility enables the early detection of potential bottlenecks, demand fluctuations, and logistical inefficiencies, enabling businesses to preemptively handle problems before they become major interruptions. Another critical capability of DTs is predictive analytics (Mihai et al., 2022). Through modeling diverse possibilities with historical and live data, companies can evaluate the potential consequences of different risk factors, such as extreme weather events or supply shortages. These simulations empower decision-makers to develop contingency plans, ensuring that businesses can adapt swiftly and maintain continuity even in adverse conditions.

Scenario simulation further enhances resilience of supply chains by enabling organizations to test and refine mitigation strategies. For example, a company can simulate the effects of a transportation delay caused by port congestion and identify alternative routes or suppliers to minimize disruptions. This proactive approach helps balance operational efficiency with adaptive capacity, reducing dependency on reactive crisis management. In addition to risk mitigation, DTs contribute to long-term optimization by identifying inefficiencies and recommending process improvements. By analyzing vast amounts of data, these models can suggest cost-saving measures, enhance sustainability efforts, and improve overall supply chain agility. This capability is particularly valuable in an era where sustainability and environmental concerns are shaping corporate and policy decisions.

Moreover, DTs enhance collaboration among supply chain stakeholders (Guo & Mantravadi, 2025). By providing a shared virtual environment, DTs enable real-time coordination between manufacturers, suppliers, and logistics providers. This interconnected ecosystem fosters better communication, reduces lead times, and enhances overall supply chain efficiency. Also, as DTs evolve, they are increasingly integrated with blockchain technology to enhance data security, traceability, and trust among participants. Despite their advantages, DT adoption is not without challenges. High costs associated with initial implementation, infrastructure upgrades, and regular maintenance expenses might be too much to afford, most notably for small- and medium-sized enterprises. Also, ensuring data accuracy and integration across disparate systems remains a significant hurdle. Cybersecurity risks also pose concerns, as the increased connectivity of supply chain operations makes them more vulnerable to cyber threats and data breaches. Overcoming these difficulties necessitates putting resources into solid cybersecurity measures, interoperability standards, and strategic partnerships with technology providers.

Given the demanding context of the semiconductor industry and the potential of DT technology, this chapter aims to provide an overview of the DT specifically designed for the semiconductor supply chain. Its purpose is twofold: first, to elucidate the fundamental architecture and core components that constitute such a DT, and second, to explore how the integration of diverse data sources enables critical functionalities and delivers tangible benefits for managing complexity and enhancing resilience. To guide this exploration, the chapter addresses the central research question: How can the integration of diverse data sources and analytical functionalities within a DT architecture enhance operational visibility, predictive capabilities, and decision-making resilience in the complex semiconductor supply chain?

Building upon the foundational concepts, this chapter conceptualizes the semiconductor supply chain DT as a dynamic, integrated virtual representation of the entire value chain, from silicon wafer sourcing to customer delivery. Powered by the continuous fusion of data from critical systems like ERP, MES, PLM, and CRM, along with planning services and external inputs, this specific DT aims to provide a synchronized, holistic view. Its core value lies in enabling real-time monitoring, predictive simulation, and intelligent optimization tailored to the unique precision, speed, and scale requirements of semiconductor manufacturing and logistics.

The remainder of this chapter is organized as follows. Section 7.2 reviews relevant literature on DTs and supply chain management. Section 7.3 details the proposed conceptual architecture for the semiconductor supply chain DT, including its core components, data hub, planning services, analytical tools, and external data integration. Subsequently, Sect. 7.4 discusses significant challenges and considerations for practical implementation, covering aspects from data integration to organizational adoption. Section 7.5 explores promising future directions for the technology. Finally, Sect. 7.6 concludes the chapter by summarizing the key findings and reinforcing the strategic importance of DTs for the semiconductor industry.

7.2 Related Work and Contribution

The concept of the DT has evolved significantly since its early conceptualization, often linked to work by Grieves in manufacturing contexts (Grieves, 2002). Essentially, a DT is a live digital model of a real-world object, operation, or system, defined by a two-way flow of data between the physical item and its virtual replica (Semeraro et al., 2021; Singh et al., 2021). This real-time or near-real-time data synchronization enables the DT to mirror the state, behavior, and context of its physical twin throughout its lifecycle. The rapid proliferation and increasing sophistication of DTs have been fueled by concurrent advancements in key enabling technologies, including the Internet of Things (IoT) for ubiquitous data sensing, AI and ML for advanced analytics and prediction, cloud computing for scalable data storage and processing, and big data analytics platforms for handling large, complex data sets (Bellavista et al., 2019). Consequently, DTs have demonstrated value across diverse sectors, including aerospace for engine health monitoring (Xiong et al., 2021), automotive for optimizing production lines and vehicle performance (Mendi, 2022; Papanagnou, 2020), healthcare for personalized medicine and hospital operations management (Sahal et al., 2022), and smart cities for urban planning and resource management (Deng et al., 2021). Across these applications, the primary objectives often revolve around enhancing operational efficiency, enabling predictive maintenance, optimizing performance, and supporting complex decision-making under uncertainty.

7.2.1 Resilience and Complexity in Modern Supply Chains

The discipline of supply chain management has grappled with the increasing complexity and fragility of globalized networks (Jones, 2023). While essential for economic efficiency (Gereffi & Lee, 2012), modern supply chains are inherently susceptible to a myriad of disruptions, encompassing natural disasters, geopolitical instability, pandemics, economic downturns, cyber threats, and logistical failures (Peck, 2006; Wagner & Bode, 2006). These vulnerabilities are often exacerbated by prevalent practices like lean manufacturing and global sourcing, which can reduce buffers and increase interdependencies. Key challenges include achieving end-to-end visibility across multiple tiers of suppliers and customers, mitigating the bullwhip effect (demand distortion amplification) (Lee et al., 1997), managing long and variable lead times, and responding agilely to volatile market demands. As a result, ensuring supply chain resilience—the capacity to anticipate, manage, and bounce back from interruptions while keeping operations running—has become an essential strategic priority (Wieland & Durach, 2021).

The semiconductor industry represents an extreme manifestation of these challenges. Its supply chain is characterized by exceptional complexity, involving hundreds of specialized processing steps, hypersensitive manufacturing environments, extremely long cycle times (months), and significant capital investment (Aelker et al., 2013). In addition, rapid technological obsolescence, intense R&D requirements, stringent quality demands bordering on zero defects, and significant geopolitical sensitivity surrounding production capacity make this sector particularly vulnerable (Chen, 2025). Traditional supply chain management tools often lack the granularity, real-time responsiveness, and predictive power needed to effectively manage such a demanding environment.

7.2.2 Digital Twins in Supply Chain Management

Recognizing the potential to address these challenges, the application of DTs to supply chains has gained traction (Bhandal et al., 2022). Existing research and applications have explored various facets of supply chain DTs. Several studies focus on optimizing logistics and transportation networks by simulating different routing strategies, carrier performances, and responses to logistical bottlenecks like port congestion (Klar et al., 2023). Others concentrate on inventory management, using DTs to simulate the impact of different inventory policies, safety stock levels, and replenishment strategies under varying demand and lead time conditions (Maheshwari & Kamble, 2022). Also, DTs are being investigated for supply chain risk assessment and resilience planning, enabling organizations to model the impact of potential disruptions (e.g., supplier failures, transportation link outages) and test the effectiveness of mitigation strategies (Ivanov & Dolgui, 2021). Frameworks for guiding DT development, like that proposed by Biller et al. (2023), offer valuable methodological starting points.

However, much of the current supply chain DT literature adopts a relatively high-level view of the supply chain or focuses on specific segments like logistics or warehousing. While valuable, these approaches often lack the deep integration with granular manufacturing operations data that is essential for industries like semiconductors. Also, comparisons with DT applications in other manufacturing sectors like automotive reveal differences; while those sectors benefit from DTs, the unique emphasis on yield prediction, complex wafer fabrication process flows managed by MESs, and the critical interplay between design (PLM), fabrication (MES), planning (ERP/APS), and customer orders (CRM) in semiconductors requires a more specialized approach (Mao et al., 2025; Singh et al., 2022). Integrating the high-volume, high-velocity data from MES and fab sensors into a supply-chain-level DT poses significant technical and conceptual challenges not always addressed in generic supply chain DT models (Negri et al., 2020).

Despite the progress in both DT technology and supply chain DT applications, a significant gap exists in the literature regarding a comprehensive, integrated, and validated architectural framework for a DT tailored specifically to the end-to-end semiconductor supply chain. Existing models often fail to adequately capture the intricate interdependencies between multitier suppliers, complex internal fabrication processes (especially MES integration), sophisticated planning requirements (including capacity, master, and sales and operations planning (S&OP)), PLM, and CRM within this specific industrial context.

This chapter addresses this gap by proposing and detailing a conceptual architecture for a semiconductor supply chain DT. Our primary contributions include the presentation of a holistic, end-to-end architectural blueprint designed explicitly for the semiconductor industry, which uniquely emphasizes the tight integration of critical systems often treated in isolation, particularly the deep linkage between shop-floor MES data and higher-level supply chain planning and execution systems (ERP, PLM, CRM, advanced planning systems). We detail the crucial role and necessary integration points for these foundational business systems alongside a suite of specialized planning and dispatching services (such as demand planning, real-time dispatching, capacity planning, and S&OP) that reflect the specific decision-making needs of semiconductor operations. The architecture also highlights the necessity of a robust supply chain data hub for harmonization and outlines the critical importance of integrating diverse external data sources like suppliers, customers, market intelligence, and IoT sensors to provide situational awareness and context. Furthermore, we clearly connect the proposed architectural components to the enablement of core functionalities—real-time monitoring, predictive simulation, and optimization—demonstrating precisely how this integrated structure delivers tangible benefits including enhanced visibility, proactive risk mitigation, improved agility, optimized resource allocation, and stringent quality assurance within the semiconductor context. By detailing these mechanisms of data and system integration, the chapter provides a concrete answer to the research question concerning how these integrations enhance operational visibility, predictive capabilities, and decision-making resilience. Ultimately, this work provides a valuable conceptual foundation and reference model for industry practitioners seeking to implement supply chain DT solutions and for researchers aiming to further investigate specific components or challenges within this domain.

7.3 Semiconductor Supply Chain Digital Twin

The semiconductor industry is characterized by high complexity, rapid technology evolution, and stringent quality requirements. A DT in this context provides a virtual replica of the entire supply chain—from raw material sourcing through manufacturing to customer delivery (Grieves, 2023). By merging data from multiple systems and external sources, the DT supports real-time monitoring, simulation, and optimization across the entire supply chain. A diagram of the DT architecture is shown in Fig. 7.1.

Core Business Systems in Semiconductor Manufacturing At the core of the DT are four crucial business systems that support a semiconductor manufacturer's operations:

- **Enterprise Resource Planning (ERP):** The ERP system tracks high-value materials such as silicon wafers, specialty chemicals, and process gases. It manages procurement, inventory, and cost analysis—ensuring that funds and materials are aligned with production needs.
- **Manufacturing Execution System (MES):** MES monitors the actual semiconductor fabrication process. It collects real-time data from fabrication tools

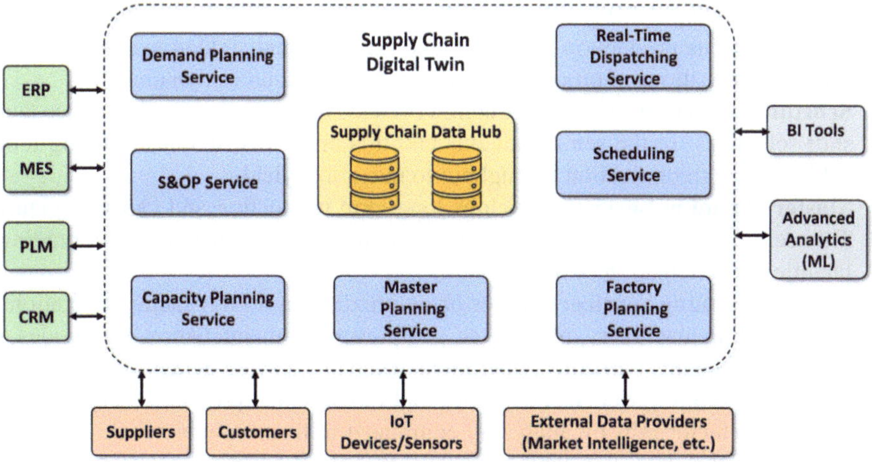

Fig. 7.1 Semiconductor supply chain digital twin

(e.g., chemical vapor deposition reactors or ion implanters), monitors process parameters, and tracks yield. This level of detail is crucial for maintaining the high quality and strict tolerances needed in chip manufacturing.
- **Product Lifecycle Management (PLM):** Semiconductor devices undergo complex design, prototyping, qualification, and mass production stages. The PLM system manages this lifecycle by tracking design iterations, version control, and compliance documents, thereby linking design data with production practices.
- **Customer Relationship Management (CRM):** CRM tools are important in managing interactions with large-scale customers who require advanced chips or microprocessors. By monitoring orders, after-sales service, and client feedback, the CRM system helps forecast demand and align production with market needs.

These systems continuously feed data into the DT. Their integration ensures that the virtual model reflects real-world production activities and business transactions accurately.

Supply Chain Data Hub At the center of the DT architecture is the supply chain data hub. This hub aggregates data from all core business systems, planning services, and external data sources. It is the centralized repository where data is stored, harmonized, and distributed. The hub is crucial for maintaining data integrity and for supporting advanced analytics and decision-making.

Planning and Dispatching Services Planning services in the semiconductor DT use the aggregated data to optimize discrete yet interdependent processes. Key services include:

- **Demand Planning Service:** Forecasts future requirements for semiconductor products using historical production data, market trends, and customer order histories. For instance, it estimates future demand for high-performance microprocessors based on emerging computing trends.
- **Real-Time Dispatching Service:** Manages the routing of in-process wafers through various production stages. It takes into account sudden machine downtimes or fluctuations in process yield to reallocate resources dynamically.
- **Scheduling Service:** Aligns production cycles with equipment maintenance and shift schedules. In the semiconductor fab, this may involve synchronizing tool calibration with production throughput to maximize yield.
- **Master Planning Service:** Develops long-term production and capacity plans. This service facilitates decisions on expanding production lines or upgrading facilities in anticipation of market shifts.
- **Factory Planning Service:** Focuses on optimizing day-to-day factory operations by scheduling machinery, manpower, and preventive maintenance. It provides a real-time mechanism to adjust operations when unexpected delays occur.
- **Capacity Planning Service:** Assesses and forecasts the production capabilities of semiconductor fabs by analyzing equipment utilization, maintenance cycles, and throughput data. For example, it determines whether critical tools such as photolithography scanners or etching systems can handle increased production demands during technology transitions.

- **Sales and Operations Planning (S&OP) Service:** Aligns market demand with production capacity by integrating sales forecasts, inventory levels, and operational data. In semiconductor manufacturing, this service ensures that the launch of new chip designs is synchronized with production schedules, thereby balancing supply with fluctuations in market demand.

Analytics and Decision Support Tools Effective analysis of the massive and diverse data sets is the backbone of a functional DT. Two categories of tools are essential:

- **Business Intelligence (BI) Tools:** These tools convert raw data into intuitive dashboards and visual reports. In semiconductor manufacturing, BI dashboards provide insights into process efficiency, yield trends, inventory status, and cost performance.
- **Advanced Analytics and Machine Learning (ML):** Leveraging historical data and real-time inputs, these tools develop predictive models. For example, ML models can forecast equipment failure in a clean room environment or identify subtle trends in defect rates during wafer fabrication—allowing preemptive adjustments.

External Data Sources To ensure the DT remains current and robust, it integrates multiple external data sources:

- **Suppliers:** Data from suppliers who provide silicon wafers, specialty chemicals, and process gases help in early detection of potential disruptions. Timely insights into supplier performance play a critical role in inventory planning.
- **Customers:** Order data and feedback from customers, such as major electronics or automotive companies, aid in refining production forecasts and aligning product specifications.
- **IoT Devices and Sensors:** In-fab sensors monitor environmental conditions (temperature, humidity, particulate levels) and equipment performance. These data points are crucial for ensuring the stringent conditions required in semiconductor manufacturing.
- **External Data Providers:** Market reports, economic indicators, and geopolitical data offer additional context that can refine demand forecasts and strategic planning.

The integration of diverse systems and services into a semiconductor supply chain DT delivers a range of benefits that fundamentally transform how operations are monitored and managed. One of the primary advantages is enhanced visibility. With real-time tracking from systems such as ERP, MES, PLM, and CRM, stakeholders obtain a consolidated, dynamic view of the entire supply chain. For example, a semiconductor manufacturer can continuously monitor inventory levels of high-purity silicon wafers, status reports from clean room sensors, and production yields from advanced MES tools. This unified visibility ensures that anomalies or deviations are quickly detected, enabling immediate corrective actions and facilitating a smooth flow from material sourcing to final product delivery.

Another critical benefit is the power of predictive analytics within the DT framework. By leveraging historical data, real-time sensor readings, and supplier performance metrics, advanced algorithms are capable of forecasting potential disruptions before they impact production. In practical terms, if a lithography machine in the fabrication facility begins to show performance degradation, predictive models can signal that a failure may be imminent. Likewise, if there is an emerging trend of delays from a chemical supplier, the system alerts managers, allowing preemptive measures such as adjusting orders or scheduling maintenance. This foresight minimizes downtime and ensures that the delicate operations of semiconductor production adhere to planned schedules.

Improved agility is a further advantage enabled by the DT. Semiconductor manufacturing is inherently dynamic, with processes that can be significantly impacted by unexpected events, such as delays in raw material shipments—a delay in the delivery of a particular grade of dopant or high-purity silicon being a common scenario. When such disruptions occur, the DT rapidly re-optimizes production schedules by reallocating resources and revising priorities based on current conditions. This capability ensures that the production line can quickly adapt to unforeseen changes, minimizing the impact on throughput and preserving the balance between demand and supply.

Process optimization is also significantly enhanced through simulation and what-if analyses provided by the DT environment. Semiconductor production involves numerous tightly coupled processes where even a slight bottleneck can have a cascading effect on overall efficiency. By virtually modeling various production scenarios, engineers can identify choke points in the workflow, such as delays occurring during thermal oxidation or chemical vapor deposition stages. For instance, by simulating an extended deposition time under different operational settings, the DT can help determine the optimal balance between throughput and process quality, which ultimately enhances overall manufacturing efficiency while reducing production costs.

Finally, continuous monitoring afforded by the DT ensures stringent quality and compliance—attributes that are particularly critical in semiconductor manufacturing. In an industry where minute deviations in process parameters can lead to significant yield loss, the ongoing surveillance of key quality indices, such as particulate levels, humidity, and temperature in controlled environments, is paramount. The DT provides a detailed, real-time record of process conditions, thereby supporting adherence to regulatory standards and facilitating audits. By ensuring constant compliance with international manufacturing standards and quality benchmarks, semiconductor companies can maintain product reliability and safeguard their reputation in highly competitive markets.

7.4 Challenges and Considerations

While the potential benefits of a semiconductor supply chain DT are substantial, its successful implementation faces significant challenges across several dimensions.

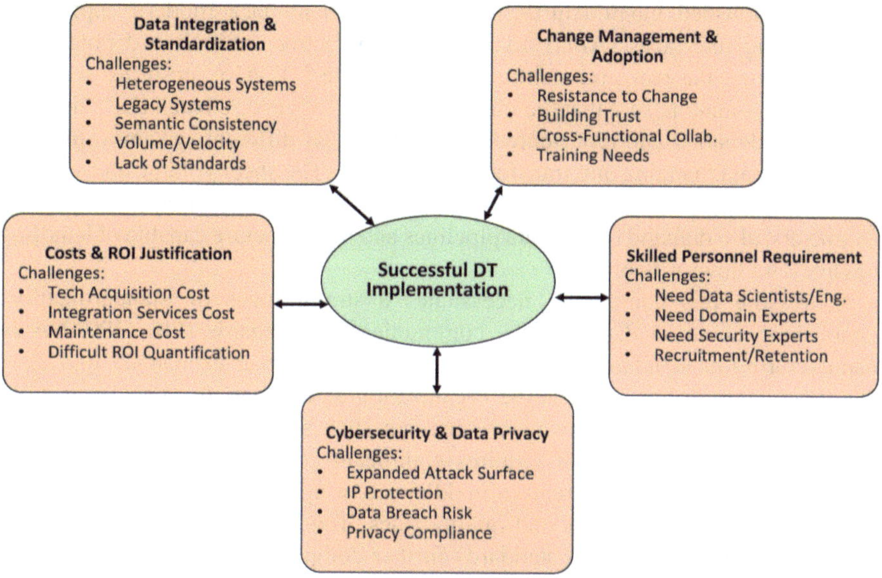

Fig. 7.2 Key challenges to successful DT implementation in the semiconductor supply chain. The diagram outlines five major categories of hurdles: ensuring robust data integration and standardization across diverse systems; justifying significant implementation costs and demonstrating return on investment (ROI); addressing cybersecurity vulnerabilities and data privacy compliance; acquiring and retaining personnel with the necessary technical and domain expertise; and managing organizational change and fostering user adoption

These challenges span technical, financial, security, personnel, and organizational aspects, as illustrated in Fig. 7.2. Addressing these proactively is crucial for realizing the full value of the DT investment.

7.4.1 Data Integration Complexity and Standardization

While the potential benefits of a semiconductor supply chain DT are substantial, its successful implementation faces significant hurdles, primarily stemming from data integration complexity. The DT relies on aggregating data from a wide array of highly diverse sources, as outlined previously. These include internal systems like ERP, MES, PLM, and CRM, real-time feeds from numerous IoT sensors within fabrication plants, and data streams from external partners such as material suppliers and logistics providers. Each of these systems often operates in distinct functional silos, utilizes different data formats, communication protocols, and data structures (schemas).

Integrating this heterogeneous data into a unified, coherent view within the DT's data hub is a major technical challenge. Legacy systems, particularly common in established manufacturing environments, may lack modern application programming interfaces (APIs), making data extraction difficult and potentially slow, hindering the real-time nature required by the DT. Also, ensuring semantic consistency—making sure that the same concept (e.g., “yield," “batch ID," “delivery date") is understood and represented uniformly across different systems—requires meticulous data mapping and transformation efforts. The sheer volume and velocity of data, especially from MES and IoT sensors capturing high-frequency process parameters, also demand robust data pipelines and infrastructure capable of handling massive data flows reliably.

Addressing this complexity necessitates a strong focus on standardization. Establishing common data models, communication standards (like OPC UA in manufacturing or standardized EDI/API formats for partner exchange), and clear data governance policies across the organization and potentially with key supply chain partners is crucial. However, achieving such standardization can be difficult and resource-intensive. It requires significant coordination among different departments, potential investment in middleware or upgrading legacy systems, and agreement on data definitions and quality rules. The lack of universally adopted, end-to-end supply chain data standards further complicates efforts, often forcing organizations to develop bespoke integration solutions. Overcoming these data integration and standardization challenges is fundamental; failure to do so can result in an incomplete, inaccurate, or lagging DT, severely limiting its effectiveness for monitoring, prediction, and optimization.

7.4.2 Implementation Costs and Return on Investment (ROI) Justification

Beyond the technical difficulties of data integration, the significant financial investment required to implement and maintain a comprehensive semiconductor supply chain DT presents another major challenge. The costs extend across multiple dimensions:

- **Technology Acquisition:** This includes licensing fees for the core DT platform, advanced analytics and AI/ML software, integration middleware, and potentially significant investments in cloud computing resources or on-premises hardware infrastructure to handle the vast data volumes and complex computations. Upgrades to existing operational systems (like MES or ERP) might also be necessary to enable effective data extraction.
- **Implementation and Integration Services:** Connecting disparate systems is complex and often requires specialized expertise, either internal or from external consultants. These integration efforts represent a substantial portion of the initial setup cost. Customizing the DT models and dashboards to accurately reflect the specific nuances of a company’s semiconductor supply chain also adds to the expense.

- **Personnel and Expertise:** Building and operating a DT necessitates skilled personnel, including data scientists, data engineers, AI/ML specialists, supply chain domain experts, and project managers. Recruiting and retaining such talent can be costly and competitive.
- **Ongoing Maintenance:** Costs do not end after implementation. Continuous software updates, platform maintenance, data quality management, model retraining and refinement, and infrastructure upkeep contribute to significant ongoing operational expenditures.

Given this substantial upfront and recurring investment, justifying the ROI becomes a critical, yet often difficult, task. While some benefits, such as reduced inventory holding costs or improved equipment utilization, may be relatively straightforward to quantify, many of the most significant advantages—like enhanced resilience against unforeseen disruptions, improved strategic decision-making agility, or increased customer satisfaction through reliable delivery—are inherently harder to translate into precise monetary figures in the short term. The complex interplay of factors influencing supply chain performance makes it challenging to isolate the exact financial contribution solely attributable to the DT. Nevertheless, developing a robust business case, potentially through pilot projects focusing on specific high-impact areas and incorporating both quantitative metrics and qualitative strategic benefits, is essential to secure executive sponsorship and ensure the project aligns with overall business objectives. Failure to adequately justify the investment can halt DT initiatives before they even begin.

7.4.3 Cybersecurity and Data Privacy Concerns

The highly interconnected nature of a semiconductor supply chain DT, linking internal OT systems like MES with enterprise IT systems (ERP, PLM, CRM) and external partner data streams, inevitably creates a significantly expanded cybersecurity attack surface. Integrating these diverse systems increases vulnerability to various threats, including unauthorized access to sensitive intellectual property (such as chip designs stored in PLM or proprietary process recipes in MES), data breaches exposing confidential supplier contracts or customer orders, and potentially malicious manipulation of data within the twin, which could lead to flawed decision-making or even physical disruption of production processes if control loops are involved. Given the high value of semiconductor IP and the critical nature of the manufacturing processes, ensuring robust cybersecurity measures is paramount. This requires putting in place several key security measures: full end-to-end data encryption, strict user access rules based on granting only minimum necessary permissions, dividing the network to keep vital systems separate, constantly watching for and detecting threats, using safe protocols for sharing information, and performing frequent security checks. Also, where personal data relating to customers or employees might be integrated, strict adherence to data privacy regulations (such as GDPR or CCPA) is mandatory, adding another layer of compliance complexity that must be addressed from the outset of the DT design.

7.4.4 Requirement for Skilled Personnel

Successfully developing, deploying, maintaining, and extracting value from a sophisticated DT demand a diverse and highly skilled workforce, presenting a significant human resource challenge. Beyond traditional IT roles, specialized expertise is essential, including data scientists capable of developing and validating complex simulation models and advanced AI/ML algorithms for predictive analytics, data engineers skilled in building robust data pipelines to handle heterogeneous data streams from sources like MES and IoT sensors, and software engineers proficient in DT platforms and integration techniques. Also, cloud architects may be needed to manage the underlying infrastructure, and cybersecurity experts are crucial for addressing the concerns outlined previously. Critically, this technical expertise must be complemented by deep domain knowledge specific to semiconductor manufacturing processes, supply chain logistics, and the business context. Recruiting and retaining individuals possessing this blend of technical and domain expertise is often difficult and costly due to high market demand, potentially impacting project timelines and budgets. Moreover, fostering effective collaboration and communication between these diverse groups—ensuring that data scientists understand the operational realities and domain experts can effectively utilize the DT's outputs—is vital for the project's success but requires conscious effort and appropriate organizational structures.

7.4.5 Change Management and Organizational Adoption

Implementing a digital twin for the semiconductor supply chain is more than simply a tech enhancement; it marks a core transformation in the established operating models, requiring significant change management efforts to ensure successful organizational adoption. Employees across various functions, from fab operators and process engineers to planners and sales managers, may exhibit resistance to adopting new tools, workflows, and data-driven decision-making processes, particularly if they perceive the technology as overly complex, opaque, or a threat to their existing roles. Building trust in the DT's data accuracy and the reliability of its analytical insights and recommendations is crucial, often necessitating efforts toward model transparency (explainable AI) and rigorous validation against real-world outcomes. The DT inherently promotes cross-functional collaboration by providing a shared, integrated view of the supply chain, potentially breaking down traditional organizational silos between R&D, manufacturing, planning, procurement, and sales. However, realizing this benefit requires fostering a collaborative culture and potentially redesigning existing business processes to leverage the system's real-time, integrated capabilities fully. Sustained executive sponsorship, a clear communication strategy articulating the vision and benefits of the DT, comprehensive user training programs, and incorporating user feedback into iterative development are all essential components of a successful change management strategy to embed the DT effectively within the organization's daily operations and decision-making culture.

7.5 Future Directions

While the current conception of the semiconductor supply chain DT offers significant advantages, ongoing technological advancements promise even greater capabilities and deeper integration in the future. The trajectory points toward more intelligent, responsive, and autonomous supply chain operations, driven primarily by progress in artificial intelligence, edge computing, and connectivity.

7.5.1 Advancements in AI/ML for Deeper Insights and Automation

Future developments in AI and ML are expected to profoundly enhance the capabilities of semiconductor supply chain DTs. Beyond current predictive analytics, future models will likely incorporate more advanced techniques such as causal inference to better understand the root causes of yield fluctuations or supply disruptions, rather than just correlating events. We anticipate the increased use of deep learning for analyzing complex, high-dimensional data from fabrication processes or unstructured data from market reports and news feeds to identify subtle patterns and emerging risks. Also, the evolution will likely move toward prescriptive analytics, where the DT not only predicts potential issues but automatically recommends optimal courses of action, potentially evaluating complex trade-offs between cost, lead time, and yield across the entire network. Reinforcement learning (RL) agents trained within the DT environment could autonomously learn optimal control policies for dynamic resource allocation, production scheduling, or inventory management, potentially achieving levels of optimization difficult for human planners (Soykan & Rabadi, 2024). The integration of natural language processing (NLP) could also enable more intuitive interaction with the DT and allow for the automated incorporation of insights from textual data sources like supplier communications or technical documentation, leading to richer context and more automated insight generation.

7.5.2 *Increased Integration with Edge Computing and Real-Time Control*

Another significant future direction involves the increased integration of DT functionalities with edge computing deployed directly within semiconductor manufacturing facilities (fabs) and logistics hubs. Processing sensor data and running localized ML models directly on edge devices, closer to the physical assets, can dramatically reduce latency compared to relying solely on centralized cloud platforms. This low-latency processing is critical for enabling near real-time feedback loops and potentially automated control actions based on DT insights. For instance, an edge-based DT component monitoring a specific piece of fabrication equipment could detect an imminent failure signature and automatically trigger a controlled shutdown or adjust process parameters on adjacent tools to mitigate the impact, minimizing yield loss or equipment damage far faster than a centralized system could react. This tighter integration between the DT's analytical capabilities and the physical operational technology (OT) layer, facilitated by edge computing and improved connectivity standards (like 5G/6G), paves the way for more responsive process control, faster anomaly detection directly on the factory floor, and ultimately, steps toward semiautonomous or fully autonomous manufacturing and logistics operations within the semiconductor supply chain. This edge intelligence can also enhance data security by processing sensitive data locally where appropriate.

7.5.3 *Expansion Across Tier-N Suppliers and Customers*

The future evolution of the semiconductor supply chain DT will likely involve a significant expansion of its scope, moving beyond the primary manufacturer's internal operations and immediate partners toward encompassing a much broader ecosystem. Currently, DTs often focus on integrating data from Tier-1 suppliers and direct customers. However, true end-to-end visibility and resilience require extending this integration deeper into the supply network, incorporating data streams from Tier-2, Tier-3, and potentially even Tier-N suppliers who provide critical raw materials, specialty chemicals, processing equipment components, or manufacturing services. Simultaneously, extending the DT's reach further downstream to include distributors, system integrators, and potentially anonymized data reflecting end-product usage or demand signals could drastically improve demand forecasting accuracy and responsiveness. Achieving this multi-enterprise network DT necessitates overcoming significant challenges related to data sharing protocols, establishing trust and data governance frameworks across disparate organizations, ensuring interoperability between diverse systems, and maintaining security across a vastly expanded network. Technologies like blockchain for secure and transparent data exchange, federated learning to train models without sharing raw data, and industry-wide data standards will be crucial for enablers for realizing this vision of a truly interconnected and collaborative digital supply chain ecosystem, providing unprecedented early warning capabilities and holistic optimization opportunities.

7.5.4 *Toward Autonomous Supply Chain Operations*

Building upon advancements in AI/ML, real-time control, and expanded network visibility, a long-term future direction points toward increasingly autonomous supply chain operations orchestrated by the DT. While current implementations primarily serve as sophisticated decision support tools providing insights and recommendations for human planners and managers, the future envisions systems capable of making and executing a wider range of operational decisions autonomously. This could manifest in various ways, such as the DT automatically adjusting production schedules across multiple fabs in response to real-time yield data or predicted demand shifts, autonomously triggering procurement orders based on predictive inventory analysis and supplier risk assessments, dynamically rerouting in-transit materials based on real-time logistics disruptions, or even self-optimizing energy consumption within manufacturing facilities based on production plans and energy pricing signals. Reaching this level of autonomy requires achieving extremely high levels of confidence in the DT's predictive models and decision algorithms, establishing robust fail-safe mechanisms and clear operational boundaries for autonomous actions, ensuring seamless integration with physical execution systems (like automated material handling systems or ERP transaction processing), and developing sophisticated governance frameworks. The ultimate goal is a self-adapting, self-optimizing supply chain ecosystem where the DT manages routine operations and responds to common disruptions autonomously, freeing human experts to focus on strategic planning, managing novel exceptions, and continuous improvement of the system itself.

7.6 Conclusion

The semiconductor industry, with its inherent operational complexities, rapid innovation cycles, and uncompromising quality standards, presents a formidable challenge for traditional supply chain management. This chapter has explored the potential of DT technology to address these challenges, proposing a tailored architecture designed to enhance resilience, visibility, and efficiency across the semiconductor supply chain.

We presented a conceptual architecture for a semiconductor supply chain DT, centered around a unifying supply chain data hub. This architecture integrates data from critical internal business systems—namely ERP for resource and material management, MES for detailed fab-level process monitoring, PLM for design and compliance tracking, and CRM for demand insights. This core is augmented by specialized planning and dispatching services (including demand, capacity, S&OP, master, factory, scheduling, and real-time dispatching) tailored for semiconductor needs, sophisticated analytics and decision support tools leveraging business intelligence and advanced AI/ML, and crucial data feeds from external sources such as suppliers, customers, IoT devices within the fabs, and market intelligence providers. This integrated structure forms the foundation for creating a dynamic, synchronized virtual replica of the physical supply chain.

The implementation of such an integrated DT yields transformative benefits specifically relevant to semiconductor manufacturing. Foremost among these is enhanced end-to-end visibility, providing stakeholders with a real-time, consolidated view across the complex network, from raw material sourcing to customer delivery. Using predictive analytics allows companies to spot possible interruptions early and take steps to lessen their impact, such as forecasting equipment failures in the fab or anticipating supplier delays, thereby minimizing costly downtime. The DT fosters significantly improved operational agility, allowing for rapid replanning and resource reallocation in response to unforeseen events. Also, it facilitates process optimization through simulation and what-if analysis, helping to identify bottlenecks and enhance manufacturing efficiency. Finally, continuous monitoring capabilities ensure stringent adherence to quality standards and regulatory compliance, which are paramount in maintaining high yields and product reliability in this demanding sector.

The semiconductor supply chain DT represents more than just a technological advancement; it is a strategic imperative for navigating the inherent complexities and intense competitive pressures of the modern semiconductor landscape. By providing unparalleled insights, predictive foresight, and enhanced control over intricate global operations, the DT empowers semiconductor companies to move beyond reactive problem-solving toward proactive, data-driven, and resilient management. Its ability to integrate disparate data sources, model complex interactions, and facilitate optimized decision-making across functional silos offers a powerful means to improve operational performance, reduce risks, shorten time to market, and ultimately secure a significant competitive advantage. As the technology continues to mature, particularly with advancements in AI, edge computing, and multi-enterprise collaboration, the DT's role as a cornerstone of efficient, agile, and resilient semiconductor supply chain operations will only become more critical.

References

Aelker, J., Bauernhansl, T., & Ehm, H. (2013). Managing complexity in supply chains: A discussion of current approaches on the example of the semiconductor industry. *Procedia Cirp, 7*, 79–84.

Banerjee, A. (2007). Global trends in supply chain planning in semiconductor industry. *Supply Chains and Firm Performance, 53*, 53–62.

Bellavista, P., Mora, A., et al. (2019). Edge cloud as an enabler for distributed AI in industrial IoT applications: the experience of the IoTwins project. In *AI&IoT@ AI* IA* (pp. 1–15).

Bhandal, R., Meriton, R., Kavanagh, R.E., & Brown, A. (2022). The application of digital twin technology in operations and supply chain management: a bibliometric review. *Supply Chain Management: An International Journal, 27*(2), 182–206.

Biller, B., Yi, J., & Biller, S. (2023). A practitioner's guide to digital twin development. In *Tutorials in Operations Research: Advancing the Frontiers of OR/MS: From Methodologies to Applications, INFORMS* (pp. 198–227).

Chen, T.C.T. (2025). Semiconductor supply chain management. In *Supply Chain Localization in the Semiconductor Industry: Rebuilding the Competitiveness and Sustainability of Semiconductor Manufacturers* (pp. 1–22). Springer.

Deng, T., Zhang, K., & Shen, Z.J.M. (2021). A systematic review of a digital twin city: A new pattern of urban governance toward smart cities. *Journal of Management Science and Engineering, 6*(2), 125–134.

Friederich, J., Francis, D.P., Lazarova-Molnar, S., & Mohamed, N. (2022). A framework for data-driven digital twins of smart manufacturing systems. *Computers in Industry, 136*, 103586.

Gereffi, G., & Lee, J. (2012). Why the world suddenly cares about global supply chains. *Journal of Supply Chain Management, 48*(3), 24–32.

Grieves, M. (2002). Product lifecycle management: driving the next generation of lean thinking. *Society of Manufacturing Engineers*, 19.

Grieves, M.W. (2023). Digital twins: past, present, and future. In *The Digital Twin* (pp. 97–121). Springer.

Guo, D., & Mantravadi, S. (2025). The role of digital twins in lean supply chain management: review and research directions. *International Journal of Production Research, 63*(5), 1851–1872.

Hicks, D.A. (1996). Evolving complexity and cost dynamics in the semiconductor industry. *IEEE Transactions on Semiconductor Manufacturing, 9*(3), 294–302.

Ivanov, D., & Dolgui, A. (2021). A digital supply chain twin for managing the disruption risks and resilience in the era of industry 4.0. *Production Planning & Control, 32*(9), 775–788.

Ivanov, D., Dolgui, A., Das, A., & Sokolov, B. (2019). Digital supply chain twins: Managing the ripple effect, resilience, and disruption risks by data-driven optimization, simulation, and visibility. *Handbook of ripple effects in the supply chain* (pp. 309–332).

Jones, S. (2023). Supply chain risk management in the era of globalization. *European Journal of Supply Chain Management, 1*(1), 11–21.

Karrila, I. (2022). Supply chain challenges in the semiconductor industry. Master's thesis, I. Karrila.

Klar, R., Fredriksson, A., & Angelakis, V. (2023). Digital twins for ports: Derived from smart city and supply chain twinning experience. *IEEE Access, 11*, 71777–71799.

Lee, H. L., Padmanabhan, V., & Whang, S. (1997). The bullwhip effect in supply chains. *Sloan Management Review, (Spring 1997)*, 93–102.

Lund, S., DC, W., & Manyika, J. (2020). *Risk, resilience, and rebalancing in global value chains*. McKinsey Global Institute, Available online at: http://dln.jaipuria.ac.in:8080/jspui/bitstream/123456789/10865/1/Risk-resilience-and-rebalancing-in-global-value-chains-full-report.pdf

Maheshwari, P., & Kamble, S. (2022). The application of supply chain digital twin to measure optimal inventory policy. *IFAC-PapersOnLine, 55*(10), 2324–2329.

Mao, R., Li, Y., Li, G., Hildre, H. P., & Zhang, H. (2025). A systematic survey of digital twin applications: Transferring knowledge from automotive and aviation to maritime industry. *IEEE Transactions on Intelligent Transportation Systems, 26*(4), 4240–4259. https://doi.org/10.1109/TITS.2025.3535593

Mendi, A.F. (2022). A digital twin case study on automotive production line. *Sensors, 22*(18), 6963.

Mihai, S., Yaqoob, M., Hung, D.V., Davis, W., Towakel, P., Raza, M., Karamanoglu, M., Barn, B., Shetve, D., Prasad, R.V., et al. (2022). Digital twins: A survey on enabling technologies, challenges, trends and future prospects. *IEEE Communications Surveys & Tutorials, 24*(4), 2255–2291.

Mönch, L., Uzsoy, R., & Fowler, J.W. (2018). A survey of semiconductor supply chain models part i: semiconductor supply chains, strategic network design, and supply chain simulation. *International Journal of Production Research, 56*(13), 4524–4545.

Negri, E., Berardi, S., Fumagalli, L., & Macchi, M. (2020). MES-integrated digital twin frameworks. *Journal of Manufacturing Systems, 56*, 58–71.

Papanagnou, C.I. (2020). A digital twin model for enhancing performance measurement in assembly lines. *Digital Twin Technologies and Smart Cities* (pp. 53–66).

Peck, H. (2006). Reconciling supply chain vulnerability, risk and supply chain management. *International Journal of Logistics: Research and Applications, 9*(2), 127–142.

Qamsane, Y., Chen, C.Y., Balta, E.C., Kao, B.C., Mohan, S., Moyne, J., Tilbury, D., Barton, K. (2019). A unified digital twin framework for real-time monitoring and evaluation of smart manufacturing systems. In *2019 IEEE 15th International Conference on Automation Science and Engineering (CASE)* (pp. 1394–1401). IEEE.

Sahal, R., Alsamhi, S.H., & Brown, K.N. (2022). Personal digital twin: a close look into the present and a step towards the future of personalised healthcare industry. *Sensors, 22*(15), 5918.

Semeraro, C., Lezoche, M., Panetto, H., & Dassisti, M. (2021). Digital twin paradigm: A systematic literature review. *Computers in Industry, 130*, 103469.

Singh, M., Fuenmayor, E., Hinchy, E.P., Qiao, Y., Murray, N., & Devine, D. (2021). Digital twin: Origin to future. *Applied System Innovation, 4*(2), 36.

Singh, M., Srivastava, R., Fuenmayor, E., Kuts, V., Qiao, Y., Murray, N., & Devine, D. (2022). Applications of digital twin across industries: A review. *Applied Sciences, 12*(11), 5727.

Soykan, B., & Rabadi, G. (2024). Optimizing humanitarian logistics with deep reinforcement learning and digital twins. In *2024 Annual Modeling and Simulation Conference (ANNSIM)* (pp 1–13). https://doi.org/10.23919/ANNSIM61499.2024.10732882

Soykan, B., Blanc, G., & Rabadi, G. (2025). A proof-of-concept digital twin for real-time simulation: Leveraging a model-based systems engineering approach. *IEEE Access*, 1–1. https://doi.org/10.1109/ACCESS.2025.3557367

Wagner, S.M., & Bode, C. (2006). An empirical investigation into supply chain vulnerability. *Journal of purchasing and supply management, 12*(6), 301–312.

Wieland, A., & Durach, C. F. (2021). Two perspectives on supply chain resilience. *Journal of Business Logistics, 42*, 315–322. https://doi.org/10.1111/jbl.12271

Xiong, M., Wang, H., Fu, Q., & Xu, Y. (2021). Digital twin–driven aero-engine intelligent predictive maintenance. *The International Journal of Advanced Manufacturing Technology, 114*(11), 3751–3761.

Part III
Methodologies and Frameworks

Chapter 8
A Framework for Supply Chain Simulation and Digital Twinning

Raid Al-Aomar, Tim Wollert, and Fabian Behrendt

Abstract This chapter presents a framework for modeling the critical functions of the supply chain and developing a simulation-based digital twin for supply chain monitoring and performance management. Digital Twin (DT), as an emerging technology, can be of great value to supply chain design, control, and informed decision-making. It can support end-to-end management of the value stream across supply chain functions (i.e., procurement, production, and distribution). The proposed framework is based on a comprehensive coverage of the supply chain using SIPOC (Suppliers–Inputs–Processes–Outputs–Customers) structure and Value Stream Digital Twinning (VSDT). It expands the scope of conventional factory-level value stream into an overall value chain using Discrete Event Simulation (DES) for supply chain modeling. The simulation serves as a flexible and relatively cheaper platform for supply chain digital twinning while being representative and supportive of timely (e.g., real-time) decision-making. The chapter outlines the supply chain modeling techniques that are essential for transforming the simulation platform from digital shadowing to digital twinning. It also discusses the technology requirements for establishing an effective simulation–physical communication and the potential integration with relevant AI applications for data analytics and machine learning.

Keywords Supply chain · Simulation · Digital twin · Value stream management · Performance management

R. Al-Aomar (✉)
German Jordanian University, Amman, Jordan
e-mail: Raid.Alaomar@gju.edu.jo

T. Wollert · F. Behrendt
Magdeburg-Stendal University of Applied Science, Magdeburg, Germany
e-mail: tim.wollert@partner.h2.de; fabian.behrendt@h2.de

G. Rabadi, B. Soykan (eds.), *Optimizing Supply Chains Through Digital Twins*,
International Series in Operations Research & Management Science 375,
https://doi.org/10.1007/978-3-032-08147-6_8

8.1 Introduction

Supply chain is an end-to-end network that enables the flow of materials and products from sources to destinations. It covers a wide scope of operations, including procurement, logistics, production, storage, and distribution. It also consists of key business elements, value, and costs. That explains the growing emphasis on effective supply chain management through the utilization of cutting-edge technologies for data analytics, optimization, and digitization. To this end, technologies such as Industry 4.0, Internet-of-Things (IoT), remote sensing, machine learning, and digital twinning are increasingly used to enable effective supply chain management across a wide range of industries.

Digital Twin (DT) is an emerging technology that can be of great value to supply chain design, control, and performance management. Digital twins can support end-to-end visibility and traceability of critical supply chain elements in addition to material flow and performance measures. To this end, DT functions as a digital replica of the physical chain that can be used for visualizing and monitoring the supply chain operations, making timely decisions to solve problems and reduce waste/cost, and predicting faults and disruptions. This can significantly reduce supply chain risks and costs and build supply chain resilience to function effectively in a dynamic and volatile environment. Other key benefits of digital twins to supply chain management include effective product distribution (e.g., shipment redirecting, inventory consolidation), transportation fleet management, shelf optimization, and efficient warehouse operations and reverse logistics.

Simulation can be used as a platform to develop the supply chain digital twin, test its functionality, and establish connectivity with the physical supply chain. Such connectivity enables the simulation to receive live data from the supply chain's ERP system for real-time visualization and measurement. Simulation outcomes can also feed the physical supply chain with live updates, adjustments, and predictions that facilitate short-term and long-term decision-making. The flexibility and the relatively low cost of simulation facilitate such functionality. Simulation can also be used for offline scenario analysis, experimentation, and performance prediction.

Although digital twin technology has been adopted across many industries, it remains underutilized in supply chains due to its relatively large scale and complex nature. Most DT applications are product-based (prototypes) or process-based within a limited scope of a workstation, an assembly line, or a facility (factory, warehouse). However, when it comes to a complex supply chain of multiple facilities in remote locations and a variety of flowing items and vehicles, exact digital replicas may not be feasible. Alternatively, simulation can serve such a role by modeling specific operations, functions, variables, and entities within the supply chain that are relevant and critical to decision-making and performance management.

This chapter presents a framework for developing a simulation-based DT of end-to-end supply chains. The framework, presented in Sect. 8.2, is based on a comprehensive coverage of the supply chain using a SIPOC (Suppliers–Inputs–

Processes–Outputs–Customers) structure. Such a structure expands the digital model of a production system to integrate the critical functions of the supply chain into a virtual platform using Discrete Event Simulation (DES) modeling. As presented in Sect. 8.3, the framework also includes a Value Stream Digital Twin (VSDT) by expanding the conventional factory-level Value Stream to cover the whole value chain. This paves the way for a new version of supply chain digital twinning that is relatively easier and cheaper to develop while being representative of its critical aspects and supportive of timely decision-making.

8.1.1 Literature Background

Current research has emphasized the need and the rationale for utilizing Digital Twins (DTs) in various industry applications. Research has also characterized the technology and business requirements of developing digital twins with emphasis on their use in visualizing and monitoring real-world systems, making timely decisions to solve and prevent problems, and predicting future performance (Tao et al., 2018). Overall, researchers agree that digital twins can greatly support smart manufacturing and can help in achieving higher efficiency, reducing costs, improving the quality, and building resilience. Soori et al. (2023) presented a comprehensive review on the structure and applications of digital twins toward smart manufacturing. Other research has highlighted DT's application to product development in terms of collaborative design and prototyping, as well as to process control in terms of monitoring and maintenance of a single machine/workstation or an assembly line (Yi et al., 2021).

Digital twins were also addressed in the context of Value Stream Mapping (VSM) as a lean management approach that is typically focused on maximizing the value delivered to customers and reducing waste across the production system (Frick & Metternich, 2022). Such a scope can be extended to a supply chain by mapping out all the steps involved in receiving materials from suppliers, processing and transporting work-in-progress and finished items, and delivering products to customers. The focus of a supply chain VSM will be on synchronizing the supply chain, reducing delays and inefficiency, and streamlining supply chain operations. For example, the order lead time can be addressed as a critical metric that directly impacts the overall supply chain efficiency, delivery speed, and customer satisfaction.

However, the dynamic nature of supply chains makes it challenging to keep the value stream maps current, especially in volatile market environments. Providing a static snapshot of the flow can only capture the state of the value stream at a single point in time (Thulasi et al., 2022). Balaji et al. (2020) highlighted several limitations of static VSM in terms of lacking up-to-date information, inaccurate representation of the actual value stream, and the difficulty of adapting the VSM to changes. An up-to-date view of the entire supply chain value stream through digital twinning has, therefore, become essential for effective monitoring, decision-making, and continuous improvement. Dynamic simulation modeling (i.e., DES) can be utilized to provide such a comprehensive view of supply chain operations (Ingalls, 2014).

The digital twinning of the supply chain VSM aims at addressing these limitations and transitioning toward a more dynamic and integrated framework that can be referred to as a digital supply chain VSM. As discussed in Frick et al. (2024), this can be attained by leveraging advanced technologies and creating a more flexible, accurate, and responsive VSM. To this end, two key challenges are considered: the automation of the VSM methodology through data integration with the physical supply chain and the expansion of the VSM structure to create a digital representation of the entire supply chain network (Frick et al., 2024). This includes processes, resources, information and material flows, and the interactions within various supply chain elements.

Most of the current work on value stream digital twinning is limited to production systems with future directions to integrate production into an overall supply chain digital twin. Simulation is typically used as a platform for dynamic VSM. For example, Wollert et al. (2024b) presented a framework for a Value Stream Digital Twin (VSDT) with a focus on designing and validating shop floor configurations. Their investigation utilized a value stream for a learning factory environment as the basis for simulating and optimizing various shop floor scenarios. Their outlook recommended considering VSDT design and requirements for a complete product life cycle or end-to-end supply chains. Further details of supply chain digital twinning can be found in Wollert et al. (2024a), Park et al. (2021), and Cimino et al. (2024).

8.2 Proposed Framework

The proposed framework builds on the work of Al-Aomar and Dweekat (2021) in simulating IoT-enabled supply chains. It can also be considered as an extension of the work of Barbosa et al. (2023) in adopting DT simulation for cycle-time analysis in a cyber-physical production system. The key objective is to utilize the DES model as a supply chain digital shadow. The model of the supply chain can be developed using any Discrete Event Simulation (DES) environment (ARENA, WITNESS, AnyLogic, FlexSim, Simul8, etc.). For example, Simul8 has a framework that is available for the development of production and supply chain digital twins (https://www.simul8.com/applications/digital-twins). AnyLogic also has a framework that integrates multiple software tools within ERP for developing a supply chain digital twin (https://www.anylogic.com/features/digital-twin/). Siemens Xcelerator also provides a digital business twin that can be utilized toward the development of an industrial Metaverse (https://xcelerator.siemens.com/).

The proposed framework is presented in Fig. 8.1. The framework reflects simulation functionality as a Digital Shadow (DS) of the physical supply chain. Using the relevant technology infrastructure, a one-way communication channel

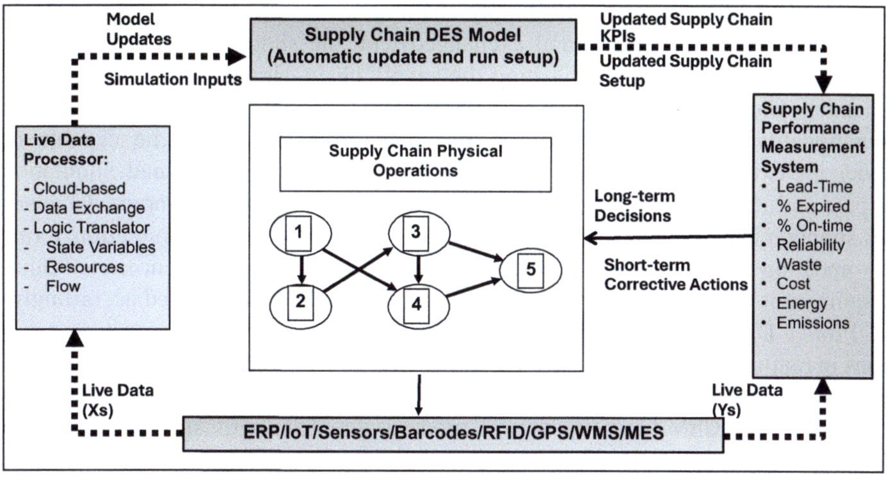

Fig. 8.1 Framework for the DT of supply chain simulation

is established (from the physical supply chain to the DES simulation model of the supply chain) for direct data exchange in real-time. This functionality can be upgraded to a bidirectional Digital Twin (DT) by mirroring the changes made in the simulation model directly into the physical elements of the supply chain. This is typically more challenging in terms of physical control (automatic or through a human-agent) and in terms of responsiveness, scale, and computational requirements. It can be included in the proposed framework through a bidirectional simulation–supply chain interaction (e.g., sensor steering, direct information flow).

As shown in Fig. 8.1, Industry 4.0 technology and the advanced infrastructure of the supply chain are utilized to establish an effective interaction between the supply chain's physical operations and the DES simulation model. An updated ERP system, as a massive database that integrates various supply chain functions, is the major source of live data for updating the simulation model. This data should be mapped with a set of input variables (Xs) that are already defined in the model and assigned to critical supply chain operations and functions. Live data should cover the three types of flow within a supply chain: material, information, and monetary. For example, the actual inventory level of a certain item can be extracted from the ERP system and used to update the relevant model variable in the simulation. Other model parameters can be assigned to ordering quantity, reorder points, and restocking level. Similar analysis of production, distribution, and logistic systems can identify more variables that are amenable for live measurement from the physical supply chain and readable by the simulation model. IoT connectivity and devices (sensors, barcodes, RFID, etc.), Manufacturing Execution Systems (MES), Warehouse Management Systems (WMS), and Global Positioning System (GPS) can be also used to generate other essential live data in relevance to supply chain entities (products, material, equipment, storage, manpower, vehicles, material handling systems, etc.).

Live data collected from the supply chain and extracted from the ERP system is expected to be processed and exchanged with the simulation model. Such an interface is essential for establishing the supply chain–simulation communication channel. Cloud-based storage and computation can be utilized to support the functionality of such a channel. A live data processor should be capable of receiving/extracting relevant data, connecting data to the predefined simulation inputs and drivers, and translating them into readable simulation inputs. Changes to supply chain logic and setup could also be reflected into simulation terms by updating model state variables, resource allocation, flow sequence, etc. Other requirements for the live update of the simulation model can be added accordingly.

From a DES perspective, an event-driven performance measurement system that runs in real-time should be enabled to effectively facilitate real-time monitoring and decision-making. The supply chain model is set up to automatically import the input data files (e.g., CSV files), update the model (variables, logic, structure), and run multiple simulation replications to estimate the supply chain performance metrics. The model should also facilitate offline experimentation to assess performance under different scenarios and conditions (market, production, supplies, logistics, etc.).

Simulation outputs (updated supply chain KPIs and setup) need to be incorporated within the supply chain performance management system. Decision-makers can greatly benefit from such a real-time dashboard of the supply chain performance under real-life conditions. Using such dashboards, decision-makers are enabled to make informed short-term decisions (e.g., corrective actions to streamline supply chain operations) as well as long-term supply chain planning decisions to deal with future events such as demand changes, supply shortages, production interruptions, etc. Integrated simulation of different supply chain functions presents an opportunity for distributed simulations in which the simulation components are spread across multiple application systems (supply chain functions). Such submodules are often connected through a network to allow for real-time interaction.

The supply chain technology infrastructure can also provide feedback to validate the simulation estimates of supply chain KPIs. Actual measurements (live data Ys) can be collected on the same set of supply chain metrics (e.g., order lead time, rejected/expired units, orders delivered on-time, delivery reliability, waste production, cost of order fulfillment, energy consumption, CO_2 emissions, etc.) and incorporated into the supply chain performance management system. Such measures can be used to validate the simulation model and confirm its credibility and can also be used to measure the actual impacts of short-term and long-term decisions on the supply chain.

8.2.1 Modeling the Supply Chain

Simulation is a computer-based model used to imitate a system's structure, functionality, and operations. It is a common engineering technique that is used in various contexts and for a wide range of applications. It has many types and forms with a wide range of industry applications from manufacturing to healthcare and service industry. Supply chains are typically simulated to understand their variable and dynamic behavior, estimate their performance under real-world conditions, design and optimize their operations, and predict their future performance. As a system-level modeling technique, DES is typically used for modeling and designing supply chains. DES is also used to model interactions among supply chain elements and to estimate and analyze the impact of implementing changes to supply chains.

Supply chain simulation expands the modeling of production/service systems to include end-to-end supply chain structure. The focus is typically on modeling material flow across the supply chain from suppliers to production, distribution, and customers. As the supply chain is mainly dealing with discrete entities (products, inventory items, shipping quantities, batches of material, production orders, etc.) and given the wide scope of supply chains, the emphasis here is on utilizing system-level DES simulation modeling as a digital model of the actual supply chain operations.

Typically, a conceptual model is first developed for the supply chain. The conceptual model should capture the network structure of the supply chain (a macro-level view of supply chain elements and interactions). The simulation process will then explore the details of supply chain elements, analyze the logical structure of their interactions, collect relevant and representative data, and translate the conceptual model into a DES simulation model of the supply chain. Once the model is verified and validated, it can be used for offline analyses of supply chain performance with findings reflected in terms of changes made to supply chain setup and structure. Figure 8.2 presents the supply chain modeling process.

The main advantage of supply chain simulation is developing a flexible platform for a holistic representation and analysis of supply chain operations and performance. This includes the modeling of suppliers and supply delivery to the production facilities, the plants' operations and inventory system, the distribution channels, warehouses, and the market order fulfillment process. The conceptual model of the supply chain network is shown in Fig. 8.3.

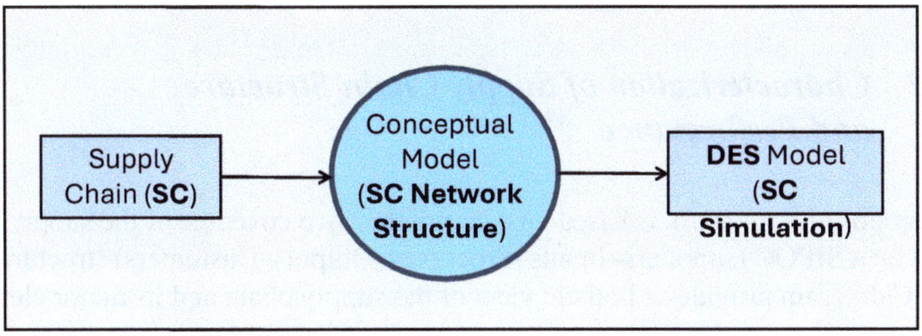

Fig. 8.2 The supply chain modeling process

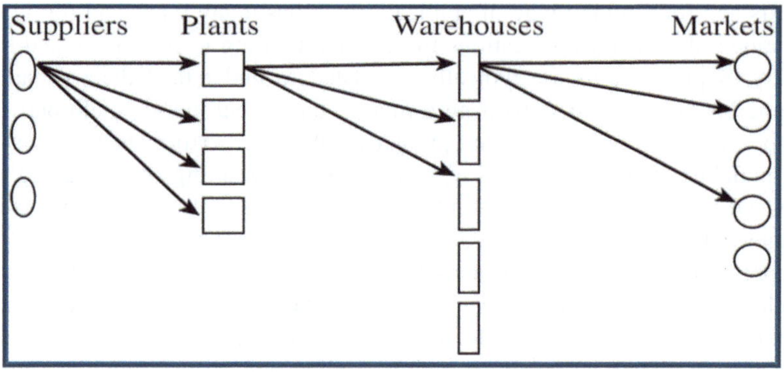

Fig. 8.3 Conceptual model of supply chain network

The main challenge of supply chain simulation is the supply chain complexity (e.g., wide scope, dynamic context, diverse functions, moving items and carriers, and multiple partners and facilities of distant locations). Practically, simulation will not be a digital replica that represents all supply chain physical elements and complex interactions. It typically acts as a digital model of the flow within the physical system, with emphasis on the critical aspects of supply chain functions. Since it is directly connected to the real world (physical system), it can be used to observe its actual behavior and to measure its performance. An effective Supply Chain Digital Twin (SCDT) can be developed by establishing a two-way interaction and communication (automatic data flow) between the supply chain simulation model and the actual supply chain operations.

The key benefit from utilizing simulation as a supply chain digital twin is the relatively low cost of simulation modeling when compared to physical modeling or to 3D scanning of the supply chain facilities and operations. Simulations are fundamentally offline models that allow analysts to experiment with a system and predict system performance. This can greatly contribute to effective supply chain design, optimization, and planning.

8.2.2 Characterization of Supply Chain Structure and Performance

The proposed framework is based on a comprehensive coverage of the supply chain based on a SIPOC (suppliers–Inputs–Processes–Outputs–Customers) structure. The SIPOC diagram provides a holistic view of the supply chain and its major elements in terms of the supplier list, supplies and inputs used in processing, main process operations, supply chain products, and key customers. To facilitate the construction of the simulation model, the SIPOC diagram can be combined with a process flow diagram to capture the structure, the flow, and interrelationships across the supply chain.

Table 8.1 Dimensions and metrics of supply chain performance

Supply chain performance dimensions and metrics				
Quality	Time	Flexibility	Cost	Sustainability
Delivery Reliability	% On-Time Delivery	Volume Flexibility	Landed Cost	CO_2 Emission
% Rejects	Order Lead Time	Mix Flexibility	Total Order Cost	Energy Consumption
% Satisfaction	Order Fill Rate	Changeover	Logistics Cost	% Recycling

Supply chain simulation and digital twinning are expected to enable accurate measurement of supply chain performance. Supply chain performance can be categorized into diverse performance dimensions such as Time, Cost, Quality, and Flexibility (Bozarth & Handfield, 2018). The proposed framework utilizes this categorization and adds to it metrics for supply chain sustainability. Time is a major aspect of supply chain performance as it determines critical metrics such as order lead time, order fill rate, and % orders delivered on time. On the other hand, cost is a key indicator of supply chain effectiveness and resilience. Key costs such as supplies landed cost, logistics cost, and total order cost are major contributors to the total cost of the supply chain.

Quality and flexibility are more dependent on the effective design of supply chain operations and the utilization of supply chain resources. Quality can be measured in terms of delivery reliability (being defect free, accurate, and complete), the % of rejected orders, and the overall service level or customer satisfaction. Flexibility is directly linked to supply chain agility, which can be a competitive advantage in today's volatile markets, under global competition.

Sustainability has also been increasingly emphasized within supply chains due to increased energy costs, strict local and global environmental regulations, and the pressure of labor and social requirements. Multiple metrics can be used for supply chain sustainability, including CO_2 emissions, energy consumption, and % recycled in reverse logistics. Table 8.1 summarizes examples of metrics that can be used in simulation and digital twins at the five dimensions of supply chain performance.

8.2.3 Simulation–Supply Chain Interaction

The simulation–supply chain interaction depends on the way in which the supply chain is modeled, the communication between the simulation model and physical supply chain, and the way in which mode simulation is utilized to configure, design, and improve the supply chain. DT is distinguished from digital models and digital shadows based on the flow of data between physical and digital objects within the supply chain. The three types of simulation interaction with the physical supply chain are:

- Simulation as a Supply Chain Digital Model (DM)
- Simulation as a Supply Chain Digital Shadow (DS)
- Simulation as a Supply Chain Digital Twin (DT)

A supply chain DM is a DES simulation model of the supply chain that is developed and operated through manual interaction with the physical system without any automatic data exchange. A supply chain DS is a DES simulation model of the supply chain that is developed and operated through one-way automatic interaction and data flow from the physical supply chain to the simulation model. Finally, a supply chain DT is a DES simulation model of the supply chain that is developed and operated to enable two-way automatic interaction and data flow. Figure 8.4 depicts the three types of simulation-based supply chain digital modeling.

Probably the supply chain Digital Model (DM) is the most common type for modeling and optimizing the supply chain. Typical applications include offline simulation-based optimization studies of the supply chain networks. There is no direct data exchange between the model and the supply chain. Supply chain data is collected and populated into the simulation model. The model is developed to represent the critical aspects of the supply chain, verified by confirming the supply chain's intended logic, and validated using actual supply chain data.

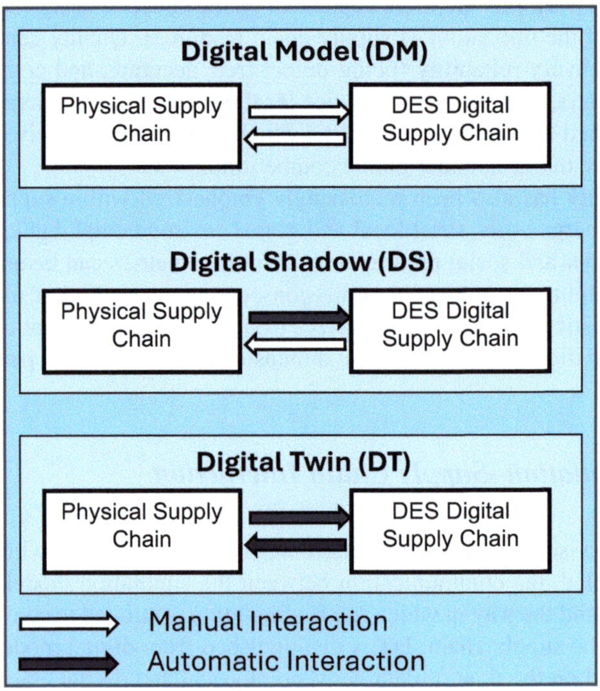

Fig. 8.4 Schematics of supply chain digital modeling

The supply chain Digital Shadow (DS) is an intermediate transition toward the full digitization of the supply chain. This refers to the one-way communication between the model and the physical supply chain (i.e., the supply chain data is transferred directly into the simulation model). The model is developed, verified, and validated in a similar manner to DM. However, the model is set up to receive and reflect the variables and performance of the supply chain in real time. This transforms the simulation into a flexible platform for monitoring supply chain performance digitally while still facilitating offline experimentation, what–if analysis, and performance prediction. Optimum parametric settings and approved structural changes that result from simulation analyses can then be adopted into the actual supply chain.

For DES to serve as a supply chain DT, the simulation model should be adapted to meet the requirements of physical system representation and interaction. A key issue in this regard is establishing direct exchange of data between the simulation and the supply chain operations in real time. As outlined in Barbosa et al. (2023), this requires the simulation model to be connected with the cyber-physical elements of the supply chain. The connection can be developed with programming languages such as R, Python, and MATLAB, or through direct simulation emulators, communication interface, and CSV files.

8.3 The Value Stream Digital Twin

This section introduces a design framework for Value Stream Digital Twins (VSDT) that represents the key elements of the supply chain. It first provides an overview of the conventional approach to Value Stream Management. The VSDT is then proposed as a representation of a holistic supply chain, followed by a discussion on the utilization of technological potential to digitize the value stream across the supply chain and the adaptation of the value stream to new challenges faced by the supply chain. The concept of the proposed VSDT is lastly validated using a Learning Experience Platform (LXP).

8.3.1 Conventional Value Stream Management

Value Stream Management is a customer-centric approach focused on optimizing the end-to-end flow of value delivered from suppliers to customers (Erlach, 2013). It originated in the 1990s and is based on lean management practices (Womack, 2003; Rother & Shook, 2018). A Value Stream Map (VSM) is a visual representation of the entire value creation process and outlines all the steps involved—both value-adding and non-value-adding. It uses defined symbols to visualize the flow of materials, information, and operations across the process, complemented by specific metrics and data to assess factors such as production lead time, degree of flow, and cycle

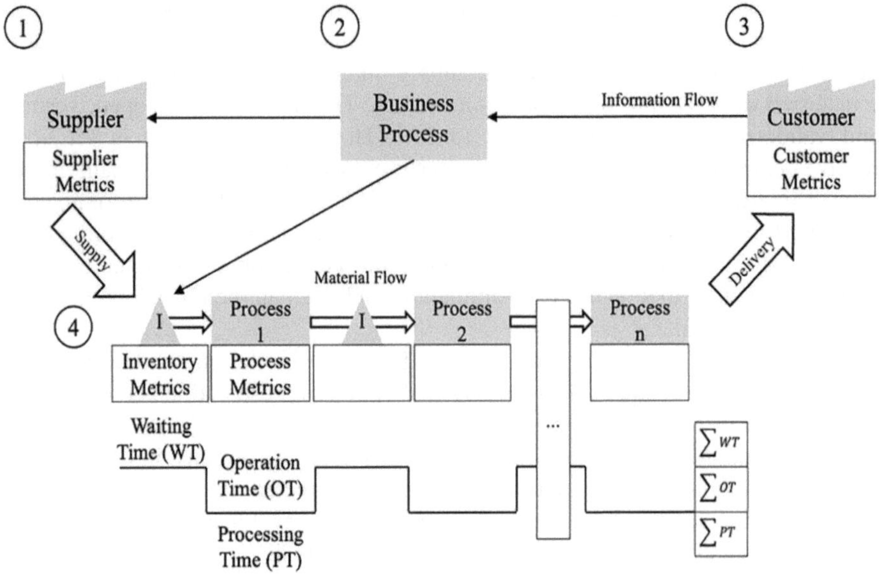

Fig. 8.5 A schematic sketch of a Value Stream Map (Erlach, 2013)

times in relation to customer takt time. In this context, the degree of flow is defined as the ratio of the sum of operation and processing times to the lead time, serving as an indicator of production dynamics. A low value correlates with inflexibility and inefficiencies, requiring improvements. The core elements of a supply chain Value Stream Map are illustrated in Fig. 8.5.

The value stream management approach begins with the documentation of the current value stream (Value Stream Mapping), whose model serves as the basis for analysis (Value Stream Analysis). Building on this, the target value stream is designed (Value Stream Design). This is followed by the phase of iterative implementation based on the Plan–Do–Check–Act (PDCA) cycle, enabling continuous improvement and optimization of the supply chain (Value Stream Planning).

Although the approach is field-tested and widely used in industry, the conventional method shows limitations in the modern environment. It is often characterized as static, inflexible, time-consuming, and imprecise. It is only a snapshot of the production that is taken based on a few measurements and averaged values (Wollert & Behrendt, 2022). In this context, various studies exist that focus on the further development of the method, considering today's requirements in production and modern design options, mainly technology driven.

8.3.2 Directions of Digital VSM

The value stream method exhibits various limitations that restrict its efficient application in a dynamic environment such as the supply chain. A key challenge is related to considering the increased digitalization in manufacturing as well as the vertical and horizontal integration within the supply chain (Magnus, 2023). Another issue could be related to the expansion of the conventional value stream model to include the information domains of costs and sustainability, by capturing domain-specific metrics related to the process (Döring et al., 2023). The conventional procedure remains unchanged in this approach.

Simulation can effectively support scenario-based investigations of different configurations in terms of Value Stream Design and concrete implementation measures. It does not address the technology challenges in the supply chain infrastructure. Results from other relevant studies can be useful for developing the VSDT. These studies range from investigating the functionality of individual technologies, such as GPS, RFID, and IoT, within the digital twin to investigating a combination of technologies, as analyzed in Wollert and Behrendt (2022). Besides technologies used at the field and control levels in the automation pyramid, operational application systems such as Manufacturing Execution Systems (MES) and Enterprise Resource Planning (ERP) serve as adequate information sources in value stream digital twinning. In addition to ERP, Supplier Relationship Management (SRM) and Customer Relationship Management (CRM) should also be considered in the value stream analysis of the supply chain.

8.3.3 Conceptual Design of the Supply Chain VSDT

A Value Stream Digital Twin (VSDT) is an advanced concept that combines the concept of Digital Twins with Value Stream Management. While a Digital Twin refers to a virtual replica of a physical object or system, linked by a bilateral automated data flow, a VSDT represents a digital twin of an entire value stream, supporting decision-makers (Frick & Metternich, 2022). This data model of the value stream is continuously updated with real-time data, offering a comprehensive and always up-to-date view. Through advanced technologies like sensors, business application systems, and data analytics, the VSDT integrates real-time data and historical data, enabling analytics to map, monitor, and optimize the value stream at every step. This digital model allows businesses to simulate and visualize business scenarios to evaluate capability and resilience, identify inefficiencies and trends or patterns, and predict potential disruptions before they occur, as well as the impact of measures on the overall performance.

The Value Stream Map merges different domains to a holistic view on the value stream, e.g., sales, production, and logistics, extended by further domains as described in the previous section. The system and process landscape of companies is always specific and requires a case-by-case assessment, as a universal approach to data processing and the design of interfaces is not feasible due to varying conditions and data conditions. For this reason, it is only possible to provide a basic layered architecture model to describe the data processing steps. An example, adapted from Wollert and Behrendt (2024), is visualized in Fig. 8.6.

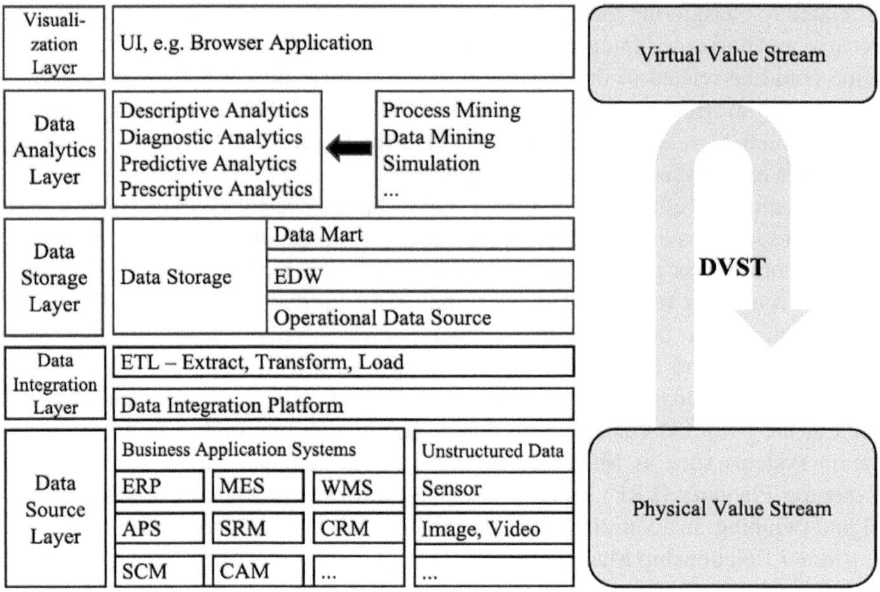

Fig. 8.6 Layered architecture model for designing VSDT (Wollert & Behrendt, 2024)

The physical value stream continuously generates data—both unstructured data and structured data stored in business application systems. These form the data sources. Using ETL (Extract, Transform, Load), the data is processed and harmonized. At the next level, the data is stored, serving as the basis for various data processing techniques in terms of data sciences. The visual representation of the digital value stream can be displayed through a user interface (UI). At this level, the virtual data model is used for monitoring, analyzing, and conducting simulation operations. Insights gained are then fed back into the physical/real value stream. This refers to adapting master data in related business application systems based on trends (e.g., planned times) as well as initiating physical changes in the value stream.

8.3.4 Enhancing the DVST for Mapping E2E Supply Chains

Based on the assumption of a supply chain as a chain of customer–supplier relationships as outlined in the SCOR model (Supply Chain Operations Reference Model; https://scor.ascm.org), a supply chain can be represented as a system of value stream, as illustrated in Fig. 8.7. Accordingly, a company acts as a customer regarding upstream supply chain stages and as a supplier regarding downstream supply chain stages. Adapted to the value stream model, the output of one value

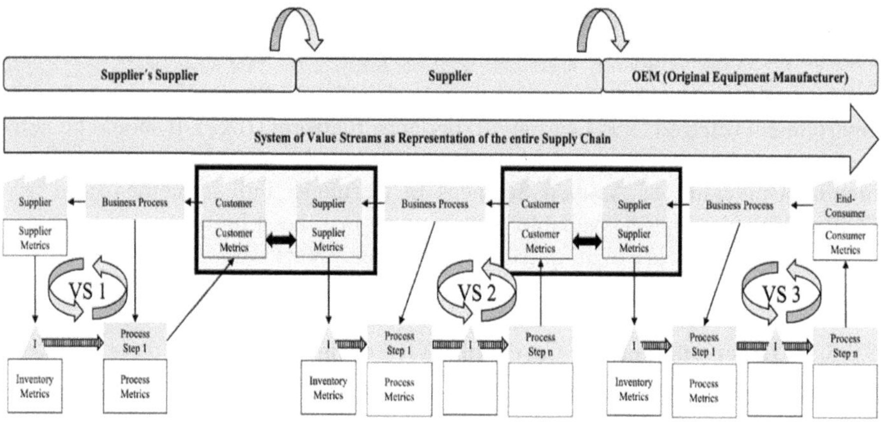

Fig. 8.7 Supply chain as a system of value streams (Wollert & Behrendt, 2024)

stream forms the input of another value stream. Following the pull principle, the KANBAN concept can be applied in a reverse direction from the end customer to the raw material supplier. In this context, each value stream can be considered both internally and externally.

A system of value streams for visualizing the entire supply chain offers several key advantages that drive operational success. It provides end-to-end visibility, allowing organizations to clearly track and trace the flow of materials and information from raw materials to the customer. This real-time transparency, in combination with historical data, helps quickly identify bottlenecks and inefficiencies, enabling timely interventions to resolve issues before they impact performance. By mapping out the value streams, businesses can enhance coordination between departments and external partners, improving communication and collaboration across the entire supply chain.

With a comprehensive view, decision-makers can make more informed, data-driven choices on inventory management, resource allocation, and production scheduling, taking the entire supply chain into account and avoiding local optimizations causing global inefficiencies. Additionally, it improves risk management by highlighting vulnerabilities and allowing proactive mitigation strategies to ensure a resilient supply chain. Focusing on value delivery at each stage aligns the supply chain with customer needs. Overall, visualizing the supply chain through an end-to-end value stream as a system of value streams supports better alignment with business goals, scalability, and long-term success. This approach offers modularity, which allows for a deep-dive examination while still maintaining integration into the overall system. Further details on the future aspects of logistics and supply chain integration can be found in Göpfert (2019).

8.3.5 *Concept Validation in Learning Experience Platform*

In this section, the concept validation is described using a Learning Factory environment referred to as Learning Experience Platform (LXP). It should be noted that this environment is used for learning and research purposes at the Magdeburg-Stendal University of Applied Sciences in Germany. Full implementation as a Digital Twin is not yet available at this point in time. Limitations and the next development stages are highlighted at the relevant points.

8.3.5.1 Description of Use Case and LXP Configuration

The LXP configuration is based on the Fischertechnik® Training Factory Industry 4.0. Its modular design enables the extension with further various models, such as a high-bay warehouse, sorting line, multi-workstation, robot, conveyor belt, and others.

The object of the investigation is the base factory. An automated storage process of raw materials into the high-bay warehouse is covered by the factory configuration, but not part of the analysis. The manufacturing process under consideration includes six operations. In the first step, materials are retrieved from the warehouse. The second and third steps simulate multi-processing operations and milling operations. In the fourth step, the workpieces are sorted by color, and in the sixth step, metadata, such as status, is written on an RFID tag at the workpiece. Finally, in the sixth step, the workpiece is prepared for shipment to the customer. A detailed description of the process and configuration is available in Behrendt et al. (2022).

The planning and control of the factory are managed through fischertechnik Cloud, a web application that covers the basic functions of Manufacturing Execution Systems (MES), including a dashboard. The production process is demand-driven, triggered by customer orders. Each workstation is connected to the cloud application, enabling real-time monitoring and control.

8.3.5.2 Architecture of the VSDT

The architecture underlying the use case, based on the general layered architecture model, is illustrated in Fig. 8.8 and is described in detail in Weigert and Behrendt (2022) and Wollert et al. (2024a).

The data source layer encompasses the entire LXP, consisting of the physical learning factory, including TXT controllers, as well as fischertechnik Cloud, which serves as the User Interface (UI) for real-time monitoring and control of the learning factory environment in real-time. The communication between fischertechnik Cloud and the workstations of the fischertechnik Training Factory Industrie 4.0 takes place via MQTT (Message Queuing Telemetry Transport).

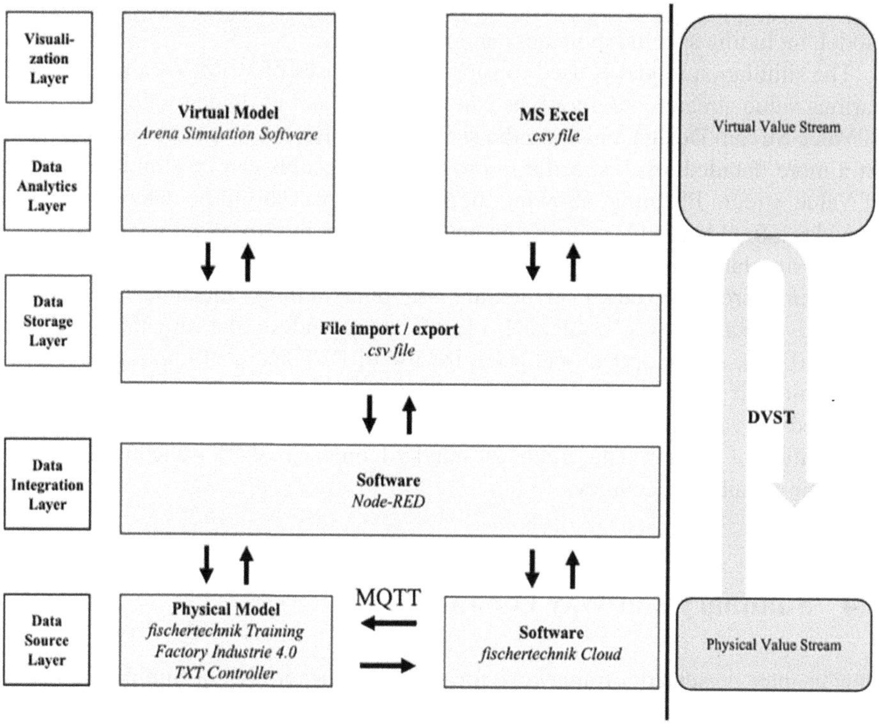

Fig. 8.8 Layered architecture model of the VSDT for concept validation (Wollert et al., 2024a)

The data integration layer consists of the software application Node-RED, which is bilaterally connected to both the physical and the virtual elements of LXP. Communication with the data source layer is also based on MQTT. While fischertechnik Cloud mainly focuses on business data, a broader range of raw data can be accessed directly through the TXT controllers. Besides import and export functions, Node-RED is used for data transformation operations.

8.3.5.3 Simulation of Supply Chain Scenarios

The focus of the study is on the application of simulation operations. The model is enriched by the continuous gathering of data from the operation of LXP. While the conventional VSM approach represents a snapshot in time, continuous data collection fundamentally enables the identification of patterns, derivation of trends, and detection of process variations, offering new fields of application. Due to the idealized environment of LXPs, disruptions must be simulated through waiting times, making them only partially representative. The gathered data is used to parameterize the simulation model, which is based on the conventional value stream model, including specific indicators and metrics.

The simulation model is used to support two phases of VSM. At a higher level, various value stream configurations can be created and evaluated within the scope of Value Stream Design, enabling the structured derivation of a target value stream. At a more detailed level, specific improvement measures can be simulated as part of Value Stream Planning, allowing their cost–benefit ratio to be determined. This provides a foundation for prioritizing and scheduling improvement measures.

The simulation of configurations considers various business scenarios, such as a sudden increase in customer demand, supplier failures, machine breakdowns, alternative technologies, bottleneck identification, and similar situations (Wollert et al., 2024b). At the operational level, the use of DVST also offers improvements. In addition to real-time monitoring, transactional data is gathered over time and analyzed for variances. Feeding the results back into the planning system enhances the quality of master data, such as standard times in work schedules, thereby increasing planning accuracy.

8.4 Summary and Way Forward

This chapter presented a framework for utilizing Discrete Event Simulation (DES) and Value Stream Management (VSM) for developing an effective Supply Chain Digital Twin (SCDT). It initially confirms the fundamental feasibility of adopting the proposed framework as a digital shadow of the physical supply chain and reveals relevant technical limitations and industry-specific challenges. It also identifies the key requirements for establishing a bidirectional simulation–physical interaction to maximize the value of supply chain digital twinning.

The chapter presents a SIPOC-based characterization of supply chain structure and outlines the Key Performance Indicators (KPIs) of SCDT that can measure and drive the continuous improvement of supply chain operations. It also discusses the process and the requirements of supply chain simulation that are essential to develop a representative and credible simulation-based SCDT. With the advancements in Industry 4.0, more data can be collected and shared with the simulation in real-time. This enables simulation as a platform for SCDT that can effectively support timely decision-making at multiple aspects of the supply chain. The presented details and examples prove the value of digital twins in supply chain management. Other examples of SCDT uses include:

- Efficient troubleshooting of problems and disruptions across the supply chain through real-time monitoring and data collection.
- Logistics and transportation management (monitoring, tracking, redirecting, etc.)
- Warehouse management (items tracking, ordering, storage, retrieval, staging, material handling, etc.)

- Effective distribution of perishable items such as dairy products to reduce their % expired and to optimize their shelf life.
- Assessing the effectiveness of reverse logistics networks through end-to-end visibility and real-time tracking of items recycling, reuse, repair, remanufacture, etc.
- Adopting green practices with a focus on real-time assessment, monitoring, and reduction of CO_2 emissions and energy consumption.
- Building resilience in supply chain operations through preventive and predictive analysis and risk reduction.
- Supporting supply chain aggregate and capacity planning through real-time data and scenario analyses under current and future conditions.
- Training and capacity building of supply chain specialists, managers, and stakeholders.

However, the proposed framework represents only a fraction of the technology and functionality requirements of supply chain digital twinning and its potential applications. On the physical side, supply chain size and complexity typically make it difficult to establish a sensory system that can effectively track and manage the flow of material, information, and monetary transactions across the supply chain. Costly infrastructure and enormous computational and data acquisition resources may be needed to provide a holistic yet detailed traceability of all supply chain entities. From the simulation side, the model cannot be a complete reflection of all supply chain elements and functions. The supply chain DES model is expected to represent a high-level view of the supply chain structure and flow and to track its most relevant variables, KPIs, and functions. Once such functionality is verified and established, the SCDT can be upscaled by integrating the details of other supply chain operations within an end-to-end SIPOC structure.

The proposed framework has multiple other limitations that can open opportunities for future research. For example, in the presented framework, variances in supply chain operations and decisions can only be simulated by varying certain input values within the value stream. Verification of the fidelity of such logic is highly needed before operational validation. Also, an individual value stream model currently exists in the framework. As the supply chain may include multiple value streams (i.e., a system of value streams), the simulation and VSM in the proposed framework need to be enhanced accordingly. The potential of data analytics and AI applications within the proposed framework will also require deeper exploration. As the supply chain has no physical learning environment, real-world disruptions are only partially replicable. For example, algorithms for real-time optimization can be used to enable timely adjustment of supply chain parameters to accommodate changing conditions, such as fluctuations in demand or supply chain disruptions. The visualization limitations of DES models are also a challenge for real-time monitoring of the supply chain's physical aspects. Several future aspects of DT application can be found in Bertoni and Bertoni (2022).

In terms of technology, future studies are expected to address the integration of supply chain simulation with ERP systems and physical IoT devices, as well as issues related to embedded sensors, software compatibility, computational capacity, connectivity, and scalability. These issues are essential for the sustainability of real-time tracking, monitoring, and mirroring. In terms of exploitation, real-time performance management can be further investigated and outlined. Optimizing forward and reverse logistic networks and inventory systems are typical examples in this regard. This also includes utilizing SCDT to facilitate collaboration with supply chain partners and the synchronization of supply chain functions, where further research is needed for developing tools and platforms that enable human interaction and seamless communication and collaboration. Finally, further research is needed to address SCDT network security and connectivity. Cybersecurity research can address the integrity of SCDT as a virtual replica of a large-scale physical supply chain.

References

Al-Aomar, R., & Dweekat, A. (2021). A simulation-based assessment of IoT-enabled supply chain of perishable dairy products. *European Journal of Industrial Engineering, 15*(6), 852–875.

Balaji, V., Venkumar, P., Sabitha, M. S., & Amuthaguka, D. (2020). DVSMS: Dynamic value stream mapping solution by applying IIoT. *Sādhanā, 45*(38).

Barbosa, V., Pinto, R., Pinheiro, J., Gonçalves, G., & Ribeiro, A.. Towards a digital twin simulation for cycle times analysis in a cyber-physical production system. In *Proceedings of the 13th international conference on simulation and modeling methodologies, technologies and applications* (SIMULTECH 2023), pp. 369–376.

Behrendt, F., Lehner, O., Rettmann, A., Schmidtke, N., & Wollert, T. (2022). Process analysis of a teaching and learning factory environment to demonstrate Industry 4.0 solutions by using the Smart Logistics Zone approach. In *2022 IEEE 6th international conference on logistics operations management (GOL)*, Strasbourg, pp. 1–10.

Bertoni, M., & Bertoni, A. (2022). Designing solutions with the product-service systems digital twin: what is now and what is next? *Computers in Industry, 138*, 103629.

Bozarth, C., & Handfield, R. (2018). *Introduction to operations and supply chain management* (5th ed.). Pearson.

Cimino, A., Longo, F., Mirabelli, G., & Solina, V. (2024). A cyclic and holistic methodology to exploit the Supply Chain Digital Twin concept towards a more resilient and sustainable future. *Cleaner Logistics and Supply Chain, 11*, 100154.

Döring, L., Wollert, T., & Behrendt, F. (2023). Evaluation of the digital transformation from VSM to Value Stream Management 4.0. In *16th International doctoral students workshop on logistics, supply chain and production management.*

Erlach, K. (2013). *Value stream design*. Springer.

Frick, N., & Metternich, J. (2022). The digital value stream twin. *Systems, 10*(4) no. 102.

Frick, N., Terwolbeck, J., Seibel, B., & Metternich, J. (2024). Design model for the digital shadow of a value stream. *Systems, 12*(20).

Göpfert, I. (2019). *Logistik der Zukunft - Logistics for the future*. Springer Fachmedien Wiesbaden.

Ingalls, R. G. (2014). Introduction to supply chain simulation. In A. Tolk, S. Y. Diallo, I. O. Ryzhov, L. Yilmaz, S. Buckley, & J. A. Miller (Eds.), *Proceedings of the 2014 winter simulation conference*.

Magnus, C. S. (2023). Smart factory mapping and design: Methodological approaches. *Production Engineering-Research and Development, 17*(5), 753–762.

Park, K. T., Son, Y. H., & Noh, S. D. (2021). The architectural framework of a cyber physical logistics system for digital-twin-based supply chain control. *International Journal of Production Research, 59*(19), 5721–5742.

Rother, M., & Shook, J. (2018). *Learning to see: Value-stream mapping to create value and eliminate muda* (1st ed.). Lean Enterprise Inst.

Soori, M., Arezoo, B., & Dastres, R. (2023). Digital twin for smart manufacturing, A review. *Sustainable Manufacturing and Service Economics, 2*, 100017.

Tao, F., Zhang, H., Liu, A., & Nee, A. Y. (2018). Digital twin in industry: state-of-the-art. *IEEE Transactions on Industrial Informatics, 15*, 2405–2415.

Thulasi, M., Faieza, A. A., Azfanizam, A. S., & Leman, Z. (2022). State of the art of dynamic value stream mapping in the manufacturing industry. *Journal of Modern Manufacturing Systems and Technology, 6*(1).

Weigert, D., & Behrendt, F. (2022). Experience Platform (LXP) to strengthen AI competence by linking simulation technologies and AI: DOI: 10.46354/i3m.2022.mas.024. In A. G. Buzzone (Ed.), *Proceedings of the 21st international conference on modelling and applied simulation, MAS*. Rende: CAL-TEK.

Wollert, T., & Behrendt, F. (2022). Automation of the manufacturing process mapping in the context of VSM by utilization of ERP Data: 10.46354/i3m.2022.mas.001. In A. G. Buzzone (Ed.), *Proceedings of the 21st international conference on modelling and applied simulation, MAS*. Rende: CAL-TEK.

Wollert, T., & Behrendt, F. (2024). Investigation of the digital twin concept to improve the value stream methodology. In M. Freitag, A. Kinra, H. Kotzab, & N. Megow (Eds.), *Lecture notes in logistics, dynamics in logistics* (pp. 431–441). Springer Nature.

Wollert, T., Henning, V., Behrendt, F., & Berndt, M. (2024a). Investigation of digital value stream twins in learning factory environments. In S. Thiede & E. Lutters (Eds.), *Lecture notes in networks and systems, learning factories of the future* (pp. 145–152). Springer Nature.

Wollert, T., Al-Aomar, R., & Behrendt, F. (2024b). Value Stream Management 4.0 – Simulating improvement measures and implementing them. *IFAC-PapersOnLine, 58*(19), 858–863.

Wollert, T., Al-Aomar, R., & Behrendt, F. (2024c). Scenario-based design and validation of shop floor configurations using an enhanced digital value stream twin. In *Proceedings of the annual industrial engineering and operations management world congress* in Detroit.

Womack, J. P. (2003). *Lean thinking: Banish waste and create wealth in your corporation*. Riverside: Free Press.

Yi, Y., Yan, Y., Liu, X., Ni, Z., Feng, J., & Liu, J. (2021). Digital twin-based smart assembly process design and application framework for complex products and its case study. *Journal of Manufacturing Systems, 58*, 94–107.

Chapter 9
"Navigating the Modern Landslide": The Agent-Based View Towards Supply Chain Immunity and Digital Twin Optimization—A Use Case with XMPro

Rajinder Bhandal and Pieter van Schalkwyk

Abstract Supply chain immunity and supply chain optimization are currently hot topics, on the back of digital twin transformation. The role of generative agents is steeped in the agent-based view; likewise, as a comparison, human behaviour is immersed in human agency. We predict that supply chains will begin to learn the same via generative agent interactions and observations in the setting of supply chains, much like human behaviour. Thus, this chapter seeks to put forward the concept of an agent-based view towards increasing supply chain optimization via a use case. There are four main themes of arrangement to this chapter, which encompass: (1) a discussion conceptualizing the notion of agent-based, supply chain immunity; (2) the application to theory relating to agentic; (3) presentation of a use case and examples; (4) a sum-up showing the application of agent-based view in supply chains via digital twins.

Keywords Agent-based · Digital twins · Generative agents · Supply chain immunity · Optimization

9.1 Introduction: The Landslide That Is Digital Movement

> Modeling what-if scenarios using agents and digital twins will become essential tool kits for proactive preparation, ahead of supply chain disruptions. This will be essential for establishing supply chain immunity in the future.
>
> **Rob Handfield, PhD,** Bank of America University Distinguished Professor of Supply Chain Management, North Carolina State University.

R. Bhandal (✉)
University of Leeds, Leeds, UK
e-mail: r.k.bhandal@leeds.ac.uk

P. van Schalkwyk
XMPro, Dallas, TX, USA

G. Rabadi, B. Soykan (eds.), *Optimizing Supply Chains Through Digital Twins*, International Series in Operations Research & Management Science 375,
https://doi.org/10.1007/978-3-032-08147-6_9

Digital transformation is moving at a fast pace, so much so that the ground is literally moving under our feet, hence the use of the term "landslide" in the title! In order to understand where we are now, we must delve into the past and recognize that this vision towards achieving supply chain optimization is built on the very foundations that still trickle through operations management!

So, what exactly is supply chain optimization and where did it all start? It is all thanks to the pioneering work of Frederick Taylor (1934) on the principles of scientific management. Taylor introduced the notion of time management in the form of studying units of time relative to a task on the factory floor, or machine shop, as it was referred to then. The earliest form of a measurement instrument was created to assess the time of completing a given task on the machine. This was known as the "observations of hand-work on machine tools" data capture method, which captured the type of operations on the machine relative to the time taken to complete the task (Taylor, 1947). Later, this method was applied to a particular part of the production line. This was in order to assess the impact of the introduction of a new machine to lift a particular part or component. That otherwise was manually carried out by the worker on the factory floor, and quite often tended to be time consuming and involve many workers. Thus, Taylor was able to assess and measure the impact of a new technique introduced on the production line, relative to time, cost, labour, and speed at which the end product was complete. Thus, the notion of Taylorism was born and became a foundation in operations management.

In fact, Henry Ford took inspiration from Taylor's Scientific Management Principles in an industrial setting, from the lens of mechanical engineering to a path towards "the same old principles in the new manufacturing" (Hounshell, 1988). Incidentally, some of the earliest work on product inventory management in the context of operations management explored methods of "improving perishable product inventory management using goal programming" (Kendall & Lee, 1980). The authors mapped out the relationship between the rate of shortages and stock-outs, managing inventory holdings, product quality, waste, and cost reduction via a goal programming technique.

Let's fast forward a bit now and consider how supply chain optimization has evolved not only to encompass operations, but also digital twin technology. In recent times, the Digital Twin Consortium has been advancing the way by providing a contemporary view on digital twins, in the form of this current definition:

> A digital twin is a virtual representation of real-world entities and processes, synchronized at a specified frequency and fidelity.
>
> - Digital twin systems transform business by accelerating holistic understanding, optimal decision-making, and effective action.
> - Digital twins use real-time and historical data to represent the past and present and simulate predicted futures.
> - Digital twins are motivated by outcomes, tailored to use cases, powered by integration, built on data, guided by domain knowledge, and implemented in IT/OT systems.
>
> Digital Twin Consortium (2020)

This chapter aims to showcase how the concept of digital twins with an agent-based approach can help enhance supply chain immunity and optimization via an XMPro use case.

This chapter is arranged in four main sections below. First, key terms in the context of the agent-based view will be defined. Second, the concept of an agent will be applied in a pragmatic manner. Third, a presentation of key examples and use cases will be shared and made available. Lastly, a discussion will be given showcasing the value of an agent-based view for supply chains.

9.2 Understanding the Key Phrases

In the first instance, it is important to unpack the notion of agency. Consequently, the concept of agency has been debated from an information systems perspective, relative to human agency and institutional interactions (Chae & Poole, 2005). Recent research by Park et al. (2023) has excelled in exploring how generative agents mimic actions and learn behaviours typical of humans. Park et al. (2023) provide a detailed definition of the term generative agent, as follows:

> Agents that draw on generative models to simulate believable human behaviour—and demonstrate that they produce believable simulacra of both individual and emergent group behaviour. Generative agents draw a wide variety of inferences about themselves, other agents, and their environment; they create daily plans that reflect their characteristics and experiences, act out those plans, react, and re-plan when appropriate; they respond when the end user changes their environment or commands them in natural language (p. 2).

The essence of agency in this context maps onto another related dimension of human behaviour! Have you heard of the concept of "being human"? The work by Margaret Archer advanced the field of sociology, thanks to her revolutionary contributions to social theory. Archer (2000) puts forward the notion that human agency is interconnected in social interactions within society, developed through multiple self and social roles. As Archer suggests, "we are aware of how our inner lives monitor our responses to external situations in which we find ourselves and indeed modify some of the circumstances to which we willingly expose ourselves, be these natural, practical or social" (p. 193). This very notion of human agency is insightful as it connects to the agent-based view in a similar tone to human behaviour, as mentioned above.

From a practical lens, another important concept that impacts various actors within a system can be understood via supply and demand, notably supply chains. Supply chain immunity as a concept was first introduced in 2023 and coined by two best-selling authors and world-renowned experts in the field of healthcare supply chains and supply chain management—Dr. Rob Handfield and Dr. Daniel Finkenstadt. Supply chain immunity is a stark wake-up call of COVID-19 past, present, and future, with valuable lessons for various stakeholders, including businesses, governments, public and private organizations, members of the public, civil servants, and humanitarian organizations. Thus, supply chain immunity per se is an important foundation for any organization to incorporate within vision and mission, in order to be future-proof.

9.3 Applications and Use Cases

Some of the earliest work in the field of operations management began in the 1980s with an effort to understand "improving perishable product inventory management using goal programming" (Kendall & Lee, 1980). The authors showcase the value of a goal programming framework, a multi-criteria tool to aid managerial decision making. Furthermore, there were various limitations relating to the model in the form of external factors. Kendall and Lee (1980) argue that "virtually all models developed by managerial decision making have neglected the unique organizational environment, bureaucratic decision processes, management style of the decision maker, and multiple conflicting nature of organizational objectives" (p. 82). Since those days, the field of operations and supply chain management has excelled, paving the way towards understanding the role of IoT and agent-based modelling in inventory and transportation efficiencies (Ding et al., 2023); digital twins and multi-modal transportation (Busse et al., 2021); and supply chain optimization with agents (Akanle & Zhang, 2008).

Table 9.1 showcases this generative agent architecture (Park et al., 2023) framework mapped onto other notable studies exploring hospital agents (Li et al., 2024) and manufacturing execution systems (Rolón & Martínez, 2012). Table 9.1 provides a summary of the way in which the MAGS agents cognitively "behave" and "operate," which can be applied to manage the vaccine supply chain through the following six various inter-related phases:

Perceive—which means to collect and acquire real-time data.
Retrieve—involves having access to historical data, events, or key knowledge.
Plan—planning takes place by using the data, historical patterns, and insights to craft out a plan or strategy of some description.
Act—this is when an optimal solution is selected, which will inform a certain desired end goal or a particular outcome.
Learn—this is when the agents add a new collection of alternatives or new knowledge towards making an optimal decision.
Reflect—this is a process of evaluation towards making effective decisions, in the long term.

9.3.1 Mapping the Agentic Approach onto the Digital Twin

At this point, you might be asking the question: "well how does the agentic approach link to the digital twin?" ... And this is a fantastic thought that can

Table 9.1 A theorization of the generative agent architecture

Adapted Generative Agent Architecture (Park et al., 2023, p.8)				
		Park et al. (2023)	Li et al. (2024)	Rolón and Martínez (2012)
INPUT	Perceive	X25 agents Small town Human behaviour "Smallville"	Hospital agents Two types: residents and medical professionals	Agent based modelling. Order agents and resource agents. Agent as a manager role. Top and bottom up.
PROCESS	Generative agent memory Memory system Retrieve Retrieved memories	Pharmacy Day in the life Social behaviour	Med-Agent-Zero method Self-evolve Simulated environment	Self-configuring. Self-healing. Self-optimizing. Self-protecting.
	Plan	Information diffusion Future sequence of action Plan memory stream	Agent based/LLM Simulate activities LLM human decision-making	Monitor. Analyse. Plan. Execute.
	Reflect	Agent's recent experience Observation from the environment	14 doctors 4 nurses Actions planned Learning during the treatment process Learning Knowledge gained by reading Medical documents Practical skills via clinical practice	Integration. Goal. Activity. Analyse. Top solution. Gantt chart.
OUTPUT	Act	Action loop Dialogue Updating plans Implementing plans Evaluation Self-knowledge	Triage Registration Consultation Medical examination	Task location. List of solutions. Execute. Check resources.

be comprehended in the following way ... There is a need to have an intelligent twin, this is because it is the ONLY way an "agent" can observe what is happening in the physical world, in keeping with Table 9.1. Thus, the digital twin actually synchronizes with the physical, as per the DTC definition: *"A digital twin is a virtual representation of real-world entities and processes, synchronized at a specified frequency and fidelity"* (Digital Twin Consortium, 2020).

If you imagine an agent, you will be aware that/or at best be mindful that it cannot ["see"] observe anything physical, thus, it has to go through a *virtual representation*, i.e., a digital twin. As a result, it is crucial that the quality of the data is high, making for a robust intelligent twin with which the agent interacts. In other words, the agent cannot "see" the supply chain (even though a supply chain per se is a non-physical asset in reality)! And thus, the agent needs to interact with the digital twin in order to see it as a physical state. The digital twin is the proxy between the agent and the physical asset, or whatever the *entity* is – it is a supply chain in this case. Therefore, the agent cannot see the supply chain; it has to look at the supply chain digital twin. Thus, a digital twin is a *physical representation* of the supply chain.

In sum, the agent needs the physical *synchronization* with the physical asset, and that way in which this is carried out is via the digital twin. The agentic concept is also dependent on *synchronized* data, which relates to the aspect [the area of the supply chain that needs attention]. Therefore, there is a need for synchronized or real-time data to identify any supply chain issues that may arise.

This relates to the problem statement or, in other words, the aspect of the supply chain that needs optimization in order to achieve supply chain "flow" (Handfield & Linton, 2022). Thus, a digital version of that supply chain is needed, as discussed in the following section below.

9.4 A Supply Chain Agentic Approach Example

9.4.1 Understanding the Supply Chain Problem and How Digital Twins Can Help

The Digital Twin Consortium® Digital Twin Capabilities Periodic Table (CPT) (Digital Twin Consortium, 2022) captures the key categories of a digital twin, which are categorized as data services; integration; intelligence; user experience; management; and trustworthiness. If we consider the category of data services, it encompasses the capability: AI model repository. Likewise, the category of intelligence includes edge AI and intelligence and prescriptive recommendations. Essentially, these capabilities are aligned with the notion of AI, analytics, and machine learning, in the overall context of a digital twin. In this chapter, we will shed more light on the value of digital twins along these lines, towards solving supply chain challenges, thus ultimately leading to supply chain optimization.

According to Ranieri et al. (2018), last-mile logistics are regarded as being the least efficient and the most complex part of a supply chain. And these inefficiencies account for around 28% of the total delivery cost! Last-mile logistics tend to be inefficient due to reasons impacting supply chains relating to cyber-attacks, weather disruptions, and a global pandemic, leading to bottlenecks (Handfield &

Finkenstadt, 2023). More specifically, cold chain logistics proved to be a complex issue during the COVID-19 pandemic, highlighting key challenges with vaccine transportation (Andoh & Yu, 2023). Thus, the main supply chain problem in this context relates to last-mile logistics—in other words, how do we improve vaccine distribution efficiency? In order to contemplate this, we first need to be mindful of what the key challenges are, as follows:

Scenario planning: "Key challenges" in vaccine distribution are:

1. **Forecasting (Handfield,** 2006)
   (a) Capacity forecasts, capacity needed, resources, capabilities.
   (b) Supply forecasting, access to data on quantity.
   (c) Country forecast, location, population, infrastructure.
2. **Inventory Management (Handfield & Linton,** 2017)
   (a) Maintain the flow of inventory.
   (b) Waste reduction management.
   (c) Speed, improving flow, supply vs. demand.
   (d) Increase movement in supply chain, over vs. under stocking.
3. **Transport and Logistics with Cold Chain Management (Handfield & Bhandal,** 2023)
   (a) Storage and handling of vaccine, maintain the correct temperature.
   (b) Transportation and handling.
   (c) Vaccine supply and distribution.
   (d) Transportation, distance, location.
4. **Last-Mile Logistics (Handfield and Linton,** 2022)
   (a) Digital dexterity.
   (b) Real-time information and data.
   (c) Supplier relationships.
   (d) Understanding supplier capabilities.
   (e) Relationship building with key suppliers.

Thus, based on the above insight, key criteria in supply chain optimization can be understood as:

1. Stock availability.
2. Effective cold chain management.
3. Transport and logistics optimization.
4. Inventory management optimization.
5. Emergency response/unexpected disruptions.
6. Quality assurance/accurate documentation/real-time data flows.
7. Ensuring compliance with regulations/public health bodies.
8. Effective teamwork and communication skills.

9.4.2 Types of Agents and Connection to Digital Twin

Typically, the agentic framework is organized around the application of large language models (LLMs) or generative AI (GenAI) running in the background when it comes to content generation. Essentially, LLMs are geared around the notion of content creation and content curation; thus, this is the main function or "role" for a content agent.

XMPro is going over and above this and is building use cases in which agents use LLMs to make decisions. If you consider a business, then the business process is rather deterministic, with regard to predicting the day-to-day flow of activities, which tend to be followed by people in the organization. Therefore, it is rather predetermined in the sense that it can be regarded as the status quo of working. This is different from that of industrial operations management and supply chain management, as what you need is data, and quite often, real-time data is highly sought-after, and thus based on the data, decisions can be made. This relates to the notion of digital twins, in particular using dashboards, underpinning the component of UX (user experience), as highlighted in the Digital Twin Capabilities Periodic Table (CPT) (Digital Twin Consortium, 2022). Thus, in this instance, the employee interacts with the data on a dashboard and can thus view certain conditions based on the operations management or supply chain objectives/goal. Thus, this is how agents contribute to the decision-making part by helping individuals in the organization make decisions around shipments, stock levels, warehouse stock control, quality, predictive maintenance, late shipments, etc. This taps into the notion of the three types of agents as characterized by XMPro, in the following way:

1. **Content Agents**
   These agents are also known as content curators, with a role focusing on researching and curating content. Content agents account for 20% of the decisions made against a backdrop of a sequential or hierarchical process. As an example, these content agents can digest various standards and requirements of a given sector and take that information to update any outdated documentation.
2. **Decision Agents**
   As the name suggests, these agents are responsible for decisions, accounting for 70% of the decisions made. In other words, 70% of use cases representing business processes are decisions, and this is in comparison to the content part, which represents 20% as noted above. The types of roles the decision agents are associated with include planning workflows, managerial duties, and quality control/quality engineering. These agents are also known as industrial experts due to their autonomous reasoning and action. For instance, these industrial experts can identify where the stock is low in a supply chain and make real-time decisions, and the same is true for predictive maintenance issues that may occur in real-time.
3. **Hybrid Agents**
   These agents account for 10% of the decisions and are a mix of content and decision-making. For example, hybrid agents can implement shipment-

related decisions, such as postponing delivery schedules, by relying on real-time data instead of paper-based documentation. The role of the hybrid agent is multifaceted, encompassing design creation, decision curation, and auditing.

With this in mind, what would an agentic "dream team" look like? The following is an attempt to theorize an agentic team with specific roles necessary for managing a vaccine supply chain, focusing on agentic capabilities aimed at meeting the specific criteria outlined below:

Who Should Be in the Team

1. Forecasting Agent
2. Cold Chain Monitoring Agent
3. Inventory Optimization Agent
4. Logistics Co-ordination Agent
5. Last Mile Delivery Agent
6. Quality Assurance Agent

9.4.3 XMPro Use Case

With all of the above in mind, let's consider how the agent-based view towards supply chain digital twin optimization plays out in the context of a use case provided by XMPro. The XMPro MAGS solution starts with the business process and creates the MAGS Team using an LLM-based wizard to do the team configuration. In essence, MAGS (Multi-Agent Generative Systems) are virtual agents that are driven by AI (artificial intelligence). These agents can be viewed as "employees" within the context of an organization. In the case of this book chapter, the focus is on supply chains. These agents tend to work independently and also have the ability to collaborate as part of a team towards optimizing efficiency and reaching a particular outcome.

Figure 9.1 captures the XMPro MAGS wizard in action for the vaccine supply chain optimization team, and when it is created in the XMPro DataStream platform, it can be edited to update communication and other parameters. It is based on real-time data that is synchronized between the data stream platform; this information is then fed into the Agents, which are configured into the Wizard.

Figure 9.2 shows that team properties can be updated and edited by duly authorized subject matter experts. There is a full audit trail of changes that records any changes made to the team or the models and parameters.

The MAGS for vaccine supply chain management consists of six specialized agents (Forecasting Agent, Cold Chain Monitoring Agent, Inventory Optimization Agent, Logistics Coordination Agent, Last Mile Delivery Agent, Quality Assurance Agent). Each agent focuses on a specific aspect of the supply chain, working together to create a comprehensive and efficient system. The MAGS for vaccine supply chain management consists of six specialized agents, and each agent focuses

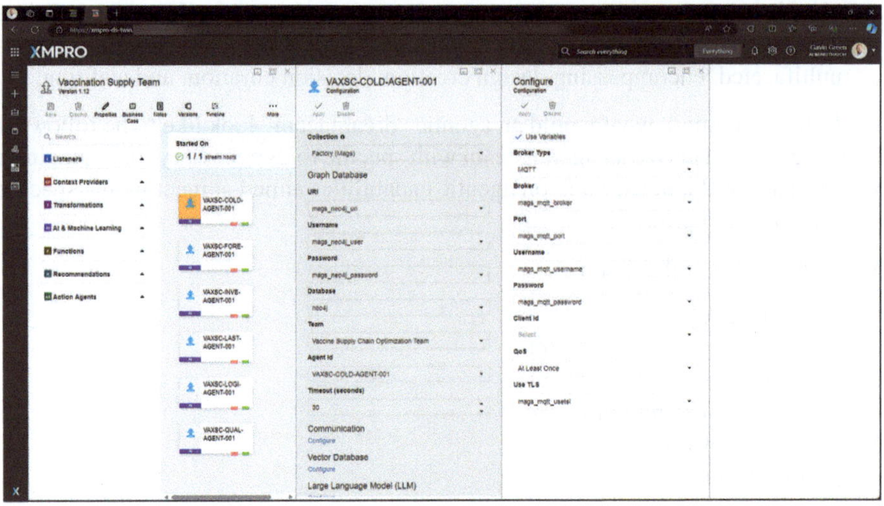

Fig. 9.1 Vaccine supply chain optimization team

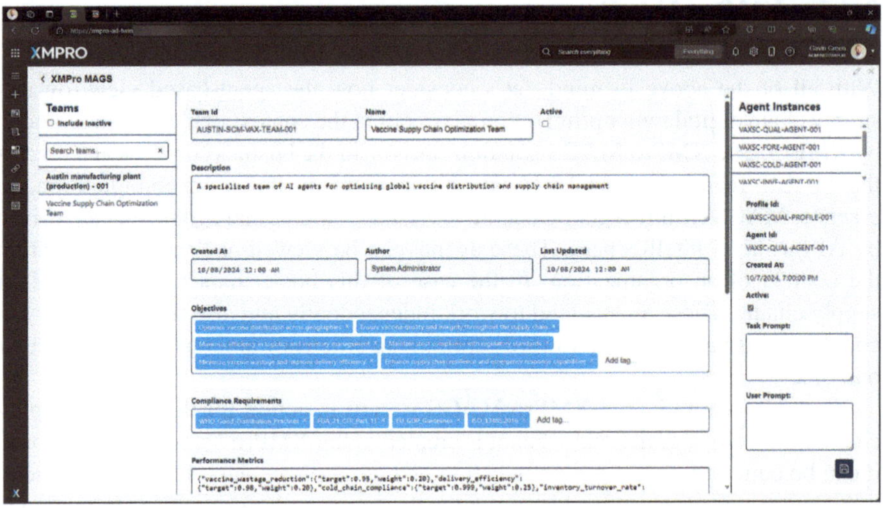

Fig. 9.2 Team properties

on a specific aspect of the supply chain, working together to create a comprehensive and efficient system. The agents are able to share insights based on real-time data towards optimizing the process of vaccine distribution from the first mile to the last mile, from the manufacturer to the end-user, in the following ways:

- Forecasting Agent—demand flows are predicated based on data such as demographics.

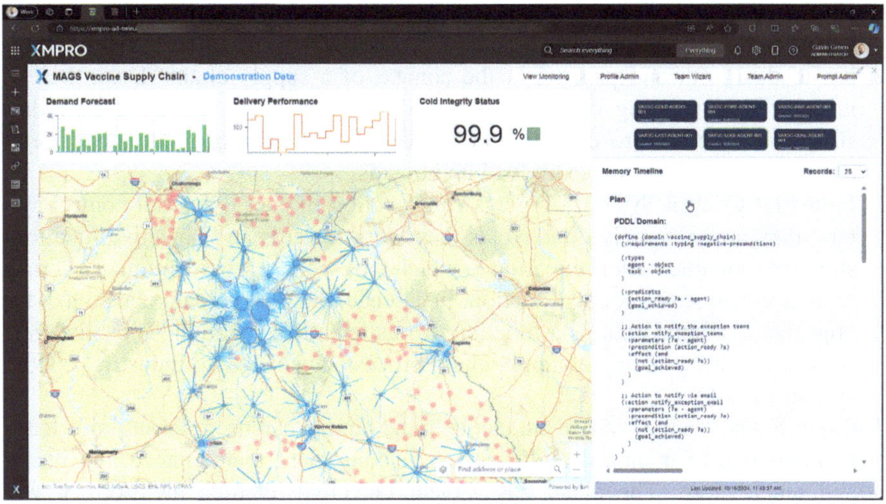

Fig. 9.3 MAGS team workflow and collaboration in vaccine supply chains

- Cold Chain Monitoring Agent—the use of technology such as IoT (internet of things) can help increase efficiency in the process of tracking the vaccine via sensors.
- Inventory Optimization Agent—these agents ensure that stock flow is consistent, and any stock waste is minimal.
- Logistics Coordination Agent—transport and logistics management towards managing transportation.
- Last Mile Delivery Agent—the role of this agent is important, and it ensures that the vaccines reach the end users in the healthcare sector at the right time, condition, and quality.
- Quality Assurance Agent—this agent helps maintain necessary regulatory compliance with regard to quality assurance.

Figure 9.3 presents the notion of team workflow and collaboration.

Collaboration is key in the context of working in teams, as this promotes trust. Thus, the effectiveness of the Multi-Agent Generative System (MAGS) in vaccine supply chain management relies on seamless collaboration among its specialized agents. Figure 9.3 outlines how the MAGS team works together to optimize the entire vaccine distribution process.

9.4.4 Benefits of the Agentic Approach

It is also important to consider the benefits of this agentic approach in the form of the following examples:

1. Cognitive Load—Agents can reduce the cognitive load in tasks. A traditional digital twin will increase the number of Alerts/Emergency Alerts that a human has to deal with and process in the context of a supply chain. Thus, there is no benefit in creating cognitive overload for humans, in terms of how they should use the data to prioritize tasks. The digital twin via generative agents can therefore reduce this extra load on humans.
2. Industrial Control System—Agents can function like an industrial control system; they are extremely effective as they work towards an "objective function" similar to a mathematical equation in which they work to try to optimize a certain scenario/ideal state/process/function. In essence, the goal is realigned so that the objective is mapped out as a function or mathematical function. Thus, an agent is effective at managing a supply chain process much like an industrial control system would.
3. Capture Knowledge—An agent can help the human capture relevant knowledge relating to user manuals, audits, machine/equipment maintenance and repair, and much more. The agent has a clear objective on this; therefore, it will communicate back to the human when there is an issue. Sometimes, this level of detail may be overlooked by humans if they are managing competing tasks.
4. Transformation—The agents are able to utilize data and advanced analysis tools in the form of AI, analytics, and machine learning, which creates positive change and transformation in operations management and manufacturing operations, as noted in MIT Machine Intelligence for Manufacturing and Operations (MIMO) (Fitzgerald, 2020).

9.5 Summary

The focus of this chapter is to shed more light on the agentic concept underpinning supply chain optimization. Thus, it is important to be able to separate out various occurrences happening in the real world vs. systems per se. In other words, the world we live in is complex and forever changing. One way to acknowledge this phenomenon is by perhaps taking a "thinking in systems" (Meadows, 2008) approach and mindset towards unpacking the reality of addressing any key challenges of an agentic-based view.

The legacy of Donella Meadows lives on; she was an author, scientist, academic, and thought leader, pioneering the notion of "systems thinking" (Meadows, 2008). Essentially, Meadows introduces the notion of how different systems within the real world, be it social, political, business, technological, or biological, to name but a few, form a "system" per se. Quite often, these systems [or system], according to Meadows (2008), are made of: (1) elements; (2) interconnections, and (3) a function or purpose (p. 11). As Meadows suggests, a system is "a set of elements or parts that is coherently organized and inter-connected in a pattern or structure that produces a characteristic set of behaviours, often classified as its "function" or "purpose"" (p.

188). This can be connected to the actual systems principles (Meadows, 2008, pp. 188–191), which can be understood as:

- Systems
- Stocks, Flows, and Dynamic Equilibrium
- Feedback Loops
- Shifting Dominance, Delays, and Oscillations
- Scenarios and Testing Models
- Constraints on Systems
- Resilience: Self-Organization, and Hierarchy
- Source of System Surprises
- Mindset and Models

With that said, this same notion can be applied as a lens to tease out any nuances in potential challenges and limitations of agentic-based view per se, and what this means for supply chains. First, let's consider the type of jobs and duties the agents may have in the following way:

A Possible Future State Scenario—Jobs and Duties of an Agent:

1. Forecasting:
   Demand forecasting; supply forecasting; mapping outbreaks across geographical area/demographics/population.
2. Inventory Management:
   Supply vs. demand—maintaining stock levels; over vs. under stocking of vaccine; vaccine category management; real-time monitoring of inventory, handling data in real time.
3. Transport and Logistics with Cold Chain Management:
   Roads, routes, locations; transportation/cold chain fleet of temperature-controlled trucks/correct/ideal temperature and storage during transition/maintain ing correct temperature.
4. Last Mile Logistics:
   Reaching the required outcomes/getting the vaccine to the right location, right time, right quality, right quantity/factoring in any unexpected distribution/transport disruptions/documentation, data updates and sharing in real time/teamwork and communication breakdown.

9.5.1 Mapping Key Challenges

If we consider the application of generative agents in supply chains, then it is important to prepare for potential challenges, which may take the form of:

Hallucinations

Hallucinations, as a concept, are traditionally associated with LLMs. Hallucinations may occur in certain instances when the data generated by the LLMs may be inaccurate or unrealistic. Therefore, this may create a situation in which the data is false, non-existent, or misinterpreted (XMPro, 2024b). This is important as it cuts across the various types of supply chain agents (as mentioned above). This is due to the fact that each agent will have a slightly different objective, as per their duty.

Trust

The Digital twin CPT (Digital Twin Consortium, 2022) captures "trustworthiness" as a category name; thus, this highlights the important role it plays within this notion of connecting the physical with the virtual world. There are particular challenges with regard to fully knowing if we can "trust" agents and if there are any biases that haven't been accounted for? The concept of trust is important as it maps to the Digital twin CPT (Digital Twin Consortium, 2022) category descriptors that include security, privacy, safety, reliability, and resilience capabilities.

9.5.2 Availability and Quality of Data

As noted above in the introduction, the DTC definition of a digital twin can be understood as "a digital twin is a virtual representation of real-world entities and processes, synchronized at a specified frequency and fidelity" thus capturing the element of "frequency and fidelity." This is important because in order for the digital twin to be accurate and operate at this level, it is reliant on having available access to high-quality data, be it historical or real-time data. This is currently a challenge within supply chains at the moment due to the nature of information sharing, access, and availability across various stakeholders and public and private organizations.

9.5.3 Potential Solutions

Here is a selection of suggested potential solutions towards dealing with some of the challenges mentioned above, as a possible guide:

1. The need to ground the flow of the agent in first principles engineering thinking and second principles engineering thinking. For example, if the stock levels decrease, should we increase stock availability? This type of data can be captured in the form of a dashboard based on the data.
2. To put this in perspective, sometimes in industry the decision-making process may involve several end-users [humans], thus this creates different perspectives across different dashboards. As a proposed alternative solution, instead of having a single dashboard for each end user. Then, how about having all of this information collectively fed into an LLM for *one* standard agent to observe, thus removing any uncertainties and potential biases in the process from say six

people with six different viewpoints. In other words, an LLM alongside agents can be grounded in order to contextualize and consolidate the data in the form of stock level data points. Thus, make predictions in which the "human" gets to determine what action is required. This way, vital data is processed without the risk of it getting missed or not accounted for.

3. Gamification can be applied to generative agents in which there is a given scenario with no rules/no code/the agents build their own memory by purely observing a situation based on real-time supply chain data. Whereby, the agents "play nicely" within the parameters, code of conduct, and ethical rules (that cannot be overridden by the agents). Thus, as a proposed solution, create a scenario as an experiment, and give the agents a platform to generate "their own" solutions to the problems that arise, as per the given supply chain issue. At the end of this experiment, all the solutions and ways in which the agents made decisions can be unpacked and evaluated by the researcher [human]. Thus, this will enable the human to see exactly how the agents have arrived at the decision. Also, this makes the process in which the agents made those decisions clear and transparent for everyone to see. In the real world, when "humans" make decisions or indeed when you try to understand "why" certain decisions were made, it is difficult to see the process, as we cannot see the mindset behind it. So, in other words, allow the agents to develop their own memory and reason to find solutions within a gamification setting for solving supply chain problems.
4. Teamwork and skills agenda for the generative agents is possibly another recommended solution. There is perhaps a need to create a pool of "problem statements" in the context of supply chains. This is towards creating a suitable environment in which the agents can learn new skills either on an individual level, and also within a teamwork setting towards dealing with key challenges, in order to find solutions. The agents will be grounded via ethical rules and grounded in their own knowledge based on supply chain data, thus reducing any potential hallucinations. The aim here is to build up the agents' skills, expertise, and experience, which closely aligns with the field of supply chain management. Perhaps, this can be viewed as "virtual HR" and professional development for agents? The irony in the term HR indicates that it is "Human" Resources, although in this context perhaps AR ["Agent" Resources] would be much more appropriate.
5. The etiquette of work and practice may well be one way to navigate through the framework of "behave," "reason," and "obligation," which encompasses the tasks and job of the agents as part of the workforce. In the context of supply chain management, it is important that the agent is focusing on the right task, and they have access to good quality data in order for the digital twin to be "intelligent" in order for the agents to be able to make sense of a particular scenario. Thus, the role of policies, obligations, permissions, and authorization can help create more flow and consistency within agent-related behaviours (XMPro, 2024a).

9.5.4 Some Final Thoughts

To sum up, there do appear to be many challenges; however, the potential solutions can offer a path towards ensuring that the agents are grounded in the knowledge to do their job. There is a need to ensure that the validation of application mechanisms (as part of the LLMs and digital twins) provides the essential oversight needed. The process of continuous monitoring is required by the "humans," ensuring that the rules of engagement are always adhered to. In the context of supply chain management, with regard to tasks, there is a need to ensure that the tasks are fine-tuned with a narrow scope. For example, the task must be specific and focus on a particular core process, as opposed to the whole job. For example, the forecast agent is solely responsible for focusing on fluctuations on supply and demand, as an example.

References

Akanle, O. M., & Zhang, D. Z. (2008). Agent-based model for optimising supply-chain configurations. *International Journal of Production Economics, 115*(2), 444–460. https://doi.org/10.1016/j.ijpe.2008.02.019

Andoh, E. A., & Yu, H. (2023). A two-stage decision-support approach for improving sustainable last-mile cold chain logistics operations of COVID-19 vaccines. *Annals of Operations Research, 328*(1), 75–105.

Archer, M. S. (2000). *Being human: The problem of agency*. Cambridge University Press.

Busse, A., Gerlach, B., Lengeling, J. C., Poschmann, P., Werner, J., & Zarnitz, S. (2021). Towards digital twins of multimodal supply chains. *Logistics, 5*(2), 25. https://doi.org/10.3390/logistics5020025

Chae, B., & Poole, M. S. (2005). The surface of emergence in systems development: Agency, institutions, and large-scale information systems. *European Journal of Information Systems, 14*(1), 19–36. https://doi.org/10.1057/palgrave.ejis.3000519

Digital Twin Consortium - The Definition Of A Digital Twin. (2020). *Digital Twin Consortium*. https://www.digitaltwinconsortium.org/hot-topics/the-definition-of-a-digital-twin/

Digital Twin Consortium- Digital Twin Capabilities Periodic Table. A Digital Twin Consortium User Guide. (2022). *Digital Twin Consortium*. 2022-03-28.

Ding, S., Ward, H., Cucurachi, S., & Tukker, A. (2023). Revealing the hidden potentials of Internet of Things (IoT)-An integrated approach using agent-based modelling and system dynamics to assess sustainable supply chain performance. *Journal of Cleaner Production, 421*, 138558. https://doi.org/10.1016/j.jclepro.2023.138558

Fitzgerald, D. (2020). *Featured Blog Post: ML in manufacturing and operations, challenges and opportunities, MIMO Presented at MIT research and development conference*. February 14, 2020. MIT. https://mimo.mit.edu/2020/02/14/ml-in-manufacturing-and-operations-challenges-and-opportunities/

Handfield, R. (2006). *Supply market intelligence: A managerial handbook for building sourcing strategies*. Auerbach Publications.

Handfield, R. & Bhandal, R. (2023). Blog: "Introduction to … Planning for the Inevitable: The Role of the Federal Supply Chain in Preparing for National Emergencies" North Carolina State University. March 19, 2023. A Guest Blog by Rajinder Bhandal Ph.D., University of Leeds. https://scm.ncsu.edu/scm-articles/article/introduction-to-planning-for-the-inevitable-the-role-of-the-federal-supply-chain-in-preparing-for-national-emergencies

Handfield, R., & Finkenstadt, D. J. (2023). *Supply chain immunity: Overcoming our nation's sourcing sickness in a post-COVID world*. Springer Nature.

Handfield, R., & Linton, T. (2017). *The LIVING supply chain: The evolving imperative of operating in real time*. Wiley.

Handfield, R., & Linton, T. (2022). *Flow: How the best supply chains thrive*. University of Toronto Press.

Hounshell, D.A. (1988). *The same old principles in the new manufacturing. From the Magazine (November 1988)*. HBR. https://hbr.org/1988/11/the-same-old-principles-in-the-new-manufacturing

Kendall, K. E., & Lee, S. M. (1980). Improving perishable product inventory management using goal programming. *Journal of Operations Management, 1*(2), 77–84. https://doi.org/10.1016/0272-6963(80)90015-7

Li, J., Wang, S., Zhang, M., Li, W., Lai, Y., Kang, X., et al. (2024). *Agent hospital: A simulacrum of hospital with evolvable medical agents*. arXiv preprint arXiv:2405.02957.

Meadows, D. H. (2008). *Thinking in systems: A primer*. Chelsea Green Publishing.

Park, J. S., O'Brien, J., Cai, C. J., Morris, M. R., Liang, P., & Bernstein, M. S. (2023). (October). Generative agents: Interactive simulacra of human behavior. In *Proceedings of the 36th annual ACM symposium on user interface software and technology* (pp. 1–22). https://dl.acm.org/doi/abs/10.1145/3586183.3606763

Ranieri, L., Digiesi, S., Silvestri, B., & Roccotelli, M. (2018). A review of last mile logistics innovations in an externalities cost reduction vision. *Sustainability, 10*(3), 782.

Rolón, M., & Martínez, E. (2012). Agent-based modeling and simulation of an autonomic manufacturing execution system. *Computers in Industry, 63*(1), 53–78.

Taylor, F. W. (1934). *The principles of scientific management*. Harper & Brothers Publishers.

Taylor, F. W. (1947). *Scientific management - Comprising shop management - The principles of scientific management and testimony before the special house committee*. Harper & Brothers Publishers.

XMPro. Deontic Principles in XMPro MAGS. (2024a). XMPro. 14-Aug-2024. https://github.com/XMPro/Multi-Agent/blob/main/docs/concepts/deontic-principles.md

XMPro. Reducing Hallucinations in Multi-Agent Generative Systems with XMPro. (2024b). XMPro. 11-Sep-2024. https://github.com/XMPro/Multi-Agent/blob/main/docs/concepts/hallucinations.md

Chapter 10
Uncertainty Quantification in Digital Twin Simulations

Bahar Biller, Canan Gunes Corlu, and Stephan R. Biller

Abstract One of the challenges of developing digital twin simulations stems from lacking full information about business process flows and characterizations of their input distributions. This chapter describes how this challenge arises in different phases of digital twin development. We present practitioners an overview of solutions to use for correctly quantifying the overall uncertainty in digital twin simulation development. We accompany the presentation with a supply chain use case.

Keywords Data-driven simulation · Digital twin · Factory twin · Input uncertainty · Risk management · Supply chain twin · Uncertainty quantification

10.1 Introduction

Digital twin arises as a critical concept for any firm on its path to digital transformation. McKinsey & Company (McKinsey&Company, 2021) estimates digital twin adoption to unlock \$6 trillion to \$13 trillion across the globe by 2030. Gartner, in their 2023 report, estimates that by 2027, 40% of large organizations will be using digital twins to increase their revenue (Gartner, 2023). Furthermore, according to the digital twin technology market report 2024–2029 (Digital-Twin-Tech, 2024), about 47% of executives understand the importance of digital twins, while 60%

B. Biller (✉)
SAS Institute, Cary, NC, USA
e-mail: Bahar.Biller@sas.com

C. Gunes Corlu
Metropolitan College, Boston University, Boston, MA, USA
e-mail: canan@bu.edu

S. R. Biller
Purdue University, West Lafayette, IN, USA
e-mail: sbiller@purdue.edu

G. Rabadi, B. Soykan (eds.), *Optimizing Supply Chains Through Digital Twins*, International Series in Operations Research & Management Science 375,
https://doi.org/10.1007/978-3-032-08147-6_10

are planning to incorporate it into their operations by 2029. The Digital Twin Consortium[1] defines digital twin as a virtual representation of real-world entities and processes, synchronized at a specified frequency and fidelity. Examples of real-world entities include components, parts, and assets, while two common examples of real-world processes come from factories and supply chains. According to the Consortium's digital twin definition, there are four main characteristics required in any digital twin: (1) a physical representation, (2) a virtual representation, (3) synchronization between physical and virtual representations at a specified frequency and fidelity, and (4) a closed feedback loop between physical and virtual representations. The first two requirements often lead to the thinking that every digital model is a digital twin. However, not every digital model is a digital twin because synchronization and learning are required for a solution to serve as a digital twin. The ability to learn and adapt leads to improved virtual models and enhancements in physical representations. In particular, the closed feedback loop informs digital models of entities and processes about the effects of solutions implemented in the physical world and enables synchronization and learning. Therefore, the closed feedback loop is the most critical aspect of any digital twin development effort.

An integrated use of digital twin-enabling technologies allows companies help their customers in various domains from aerospace and defense, healthcare, and manufacturing to retail and consumer goods, transportation, and energy and utilities. Biller et al. (2023) present a collection of industrial use cases where the benefits of the digital twin technology range from performance monitoring and data accuracy enhancements to increased financial outcomes, service improvements, and increased resilience. More generally, there are three major benefits of digital twin development. First, digital twins help understand what did happen, predict what may happen, and prescribe actions to address future problems before they happen. Second, they enable the assessment of the impact of operational policy, design, and investment decisions in a virtual laboratory. Third, digital twins help stress test the system under consideration and identify best courses of action to take when faced with disruptions. These benefits are the products of integrated use of a wide range of digital twin-enabling technologies ranging from artificial intelligence (AI), simulation, forecasting and optimization to econometrics, visual analytics, and streaming analytics. In this chapter, the focus is on the quantification of "zone-of-uncertainty" that arises when simulation is utilized as a digital twin-enabling technology.

As the industrial use case, we consider a factory twin built as a critical component of a supply chain network of suppliers, plants, warehouses, and customers. Factory twin, a type of process twin illustrated in Fig. 10.1, captures a manufacturing process flow that entities follow in a factory as they travel from one machine to another and

[1] The Digital Twin Consortium was founded in 2020 with the mission to bring corporations, innovators, academia, and governments together to accelerate the development, adoption, and widespread use of the twin technology for digital transformation.

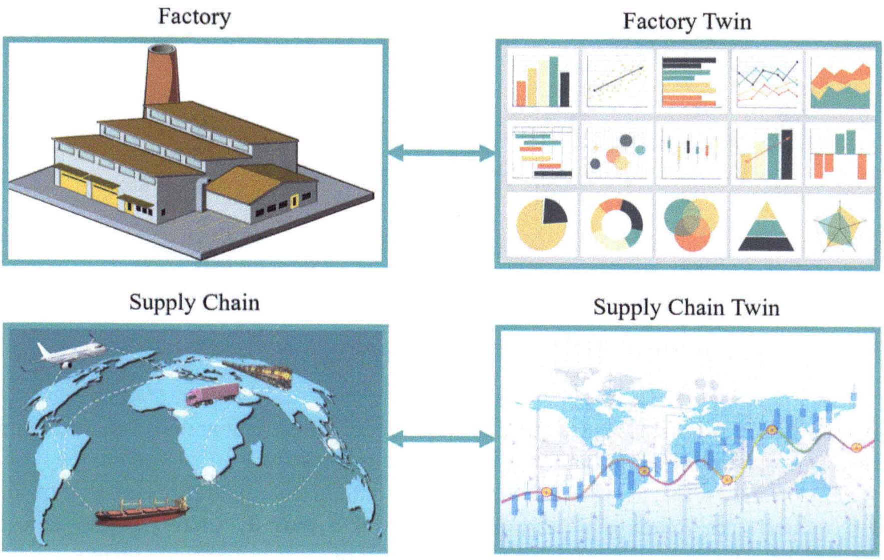

Fig. 10.1 Illustrations of factory and supply chain twins

are gradually transformed from raw materials into finished goods. In the remainder of this chapter, we restrict our focus to zone-of-uncertainty as a concept that arises during digital twin simulation development. We will not delve into the details of digital twin development and the descriptions of frameworks that can be utilized for building digital twins. However, we refer the reader to Biller et al. (2022), Biller et al. (2023), and Biller et al. (2023) for background information, where the authors discuss several different development frameworks. Biller and Biller (2023) note that the digital twin life cycle also includes digital twins of product, usage, and service. In this chapter, however, we restrict our focus to factory and supply chain digital twins. In the remainder of the section, we provide high-level overviews of the phases of digital twin development and the concept of zone-of-uncertainty as the digital twin development challenge addressed in this chapter.

10.1.1 Phases of Digital Twin Development

Our goal is to give readers an overview of the available capability to address zone-of-uncertainty within the context of building simulation-based factory and supply chain digital twins. We note that not every simulation model would qualify as a digital twin: Synchronization and learning distinguish a stand-alone simulation model from a digital twin simulation. Digital twins, by definition, evolve over the life cycles of entities and processes, going through three main development phases: offline model development, real-time synchronization, and online learning. Figure 10.2 illustrates

Fig. 10.2 Digital twin development phases

that creating a digital twin starts with offline model development and continues with real-time synchronization, online learning, and control.

The offline factory model development first builds a virtual representation of the factory and then validates it with static data representing the history of the factory operations. The validated model is used to replay history and provide insights about optimizing factory performance. In this phase of the development, the first two digital twin required characteristics are met, and the offline learning function of the development is completed. As shown in Fig. 10.2, the next phase of the digital twin development is real-time synchronization. This is when the model parameters are updated, and the factory simulation is calibrated with the most recent data at a suitable frequency. Consequently, the factory and its digital representation are synchronized. It is important to note that the synchronization frequency is based on the timeline of decision-making and the requirements of learning and adaption. For example, inventory policies might be adjusted on a weekly or monthly basis, while the preferred unit of time for emergency maintenance decisions could be a minute. The final phase of the digital twin development is online learning, involving monitoring factory operations, tracking the past, predicting future factory performance, and identifying the best course of action in response to performance degradation or disruptive events. Thus, this is when the factory twin is expected to provide enhanced visibility into the future, enable playing operational what-if games, and offer decision support for optimal performance. Simulation plays a key role in equipping a digital twin with these capabilities. Because digital twin development is motivated to achieve meaningful business outcomes that can be clearly stated and objectively measured, it is of critical importance to accurately predict the factory key performance indicators (KPIs) and accompany each prediction with a zone-of-uncertainty, which correctly represents the overall uncertainty, at a chosen confidence level. In the remainder of the chapter, we choose discrete-event simulation as the digital twin-enabling technology of interest and address the topic of zone-of-uncertainty for different phases of digital twin development.

10.1.2 Overall Uncertainty Quantification in Digital Twin Development

Details of factory and supply chain operations discussed in this chapter are motivated by a semiconductor manufacturing use case introduced in Biller et al. (2019). The decision timeline of the use case starts with strategic equipment portfolio selection. It moves toward operational management of equipment, inventory, process, and people to deliver a unit cost of production and a level of product quality to compete in the market successfully. Biller et al. (2019) present a detailed discussion of all these aspects of factory operations management. Our goal is, however, to discuss the quantification of overall uncertainty in factory KPI prediction.

The concept of zone-of-uncertainty arises due to the lack of complete characterization of the simulation input distribution functions, turning the correct quantification of the overall uncertainty in the simulation outputs into a challenging task. This is the case because we must define and capture an additional layer of uncertainty that we refer to in this chapter as the input uncertainty. During the offline modeling phase of digital twin development, the source of input uncertainty is a set of assumptions made to solve the equipment portfolio selection problem. In Sect. 10.4, we first discuss the characterization of assumption uncertainty in a stochastic simulation and describe the quantification of the corresponding additional layer of uncertainty in simulation outputs. We accompany this presentation with the implementation details and the learnings acquired by following this practice for silicon carbide fabrication design and operational policy selection at the New York Power Electronics Manufacturing Consortium facility (Biller et al., 2019).

Over the course of any project that starts with strategic investment planning, the source of input uncertainty eventually changes from only an assumption set to the availability of limited historical data collected during process verification and product qualification. The issue of only limited data availability also arises during the online learning phase when we track the past and monitor factory operations. Although it might be expected to accumulate sufficiently long datasets for simulation input modeling, existing products and processes naturally retire while new products and processes are introduced. Thus, it remains important to quantify the effect of input uncertainty in stochastic simulations supporting digital twin development. We discuss this problem within the context of factory KPI prediction and inventory control as the timeline of the use case of interest moves from strategic investment to tactical planning and eventually to operational planning.

It is a well-known practice to account for stochastic uncertainty in a stochastic simulation (Law, 2024). Specifically, the stochastic uncertainty is the inherent variability in the simulation outputs due to the dependence of the factory KPIs on the simulation's input models. We use the number of replications to control its contributions to the variability in the simulation outputs. Our goal is, however, to provide a detailed description of the input uncertainty. To do that, we consider the processing time of a part at a machine because it is one of the input random variables to be modeled in a factory simulation. Suppose that the true distribution to characterize the randomness of this processing-time variable is an exponential distribution with a mean value of 30 minutes per part, but this value is unknown to us. We use β to denote the mean processing-time parameter in hours, note that $\beta = 0.5$ is unknown to us, and assume the availability of only a limited amount of data of length N. What is important to recognize here is that this limited dataset represents a single realization (i.e., a single sample path) generated by the true demand distribution (i.e., the exponential distribution with a 30-minute mean), and we must use this length-N history to predict the mean processing time β. A way of obtaining an estimate $\hat{\beta}$ of β is via the maximum likelihood estimation method. If we define x_n, $n = 1, 2, \ldots, N$ for the historical dataset of the processing-time random variable X, then we characterize the maximum likelihood estimate (MLE) of β as

$$\hat{\beta} = \bar{x}_N = \frac{1}{N}\sum_{n=1}^{N} x_n.$$

It is well known that $\hat{\beta}$ is often strongly consistent and asymptotically normal as $N \to \infty$. Our focus is, however, on the case of working with limited data, thus relatively smaller values of N. At the heart of the challenge discussed in this chapter lies the fact that $\bar{x}_N$ is a random variable. Our goal is to address the representation of the uncertainty in $\bar{x}_N$ and in any other relevant stochastic input characterization in a factory simulation built to support digital twin development. We illustrate this uncertainty quantification challenge in Fig. 10.3.

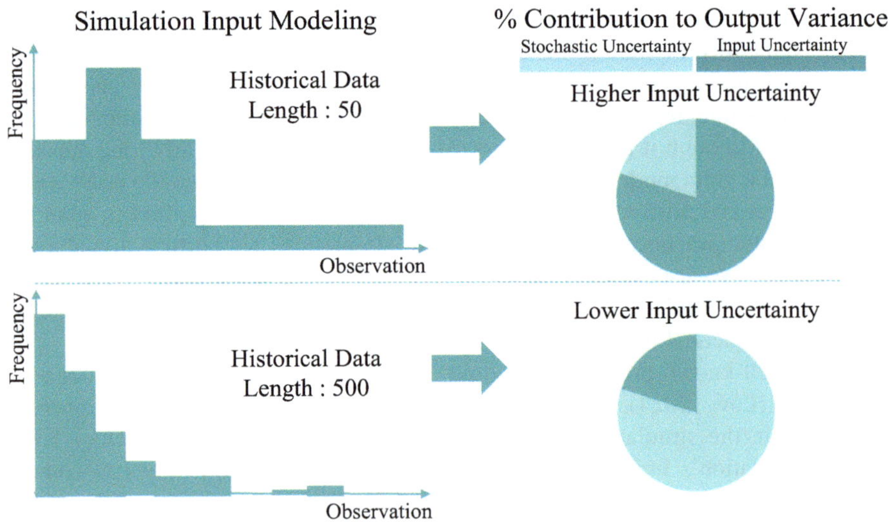

Fig. 10.3 Illustration of uncertainty quantification in the KPI output variance

Another example of a random variable that almost always arises in any supply chain digital twin simulation is product demand. Often, the parameters of demand distributions are unknown, and they must be estimated from a single realization of the demand history. Suppose that ten demand realizations have been generated from an exponentially distributed demand process with a mean value of 50 units. Thus, the exponential distribution with a mean of 50 units serves as the true distribution. However, this demand generation process is unknown to us as we develop the supply chain digital twin, and we have access only to those ten observations. Using an input-modeling software package, let us assume that exponential distribution with an estimated mean of 40 units is identified as the best fit. However, if we replace this demand series with another set of ten observations generated from the same exponential distribution, then this time the mean demand parameter estimate can be identified as 60. Every time we independently generate ten observations from the same exponential distribution, we would obtain different values for the demand parameter estimates. This means that when the length of the demand history is short, there is uncertainty around the demand parameter estimates, and it is the aim of this chapter to discuss the characterization of this type of demand (input) uncertainty. We caution practitioners against the danger of sequentially estimating the demand parameters first and then solving the inventory optimization problem based on these estimates. This is the case because such an approach would ignore the demand parameter uncertainty and, therefore, the uncertainty around the inventory-target estimate. We further elaborate on this topic in Sect. 10.4.3.

We aim to equip practitioners with the knowledge that can be utilized to correctly quantify input uncertainty during offline modeling and online learning phases of digital twin development. We organize the remainder of this chapter as follows. Section 10.2 reviews the existing body of knowledge about digital twins and stochastic simulations driven by limited information and data. Section 10.3 provides a formal description of the concept of zone-of-uncertainty. Section 10.4 discusses how the problem defined in Sect. 10.3 has been addressed in an industrial use case via the utilization of selective methods reviewed in Sect. 10.2. We conclude our presentation in Sect. 10.5 with a brief discussion about the outstanding challenges of input uncertainty quantification in simulation-based digital twin development.

10.2 Literature and Contribution

In Sect. 10.2.1, we provide selected sources of knowledge for digital twins. In Sect. 10.2.2, we switch our focus to simulation digital twins. In Sect. 10.2.3, we review the literature on quantification of simulation input uncertainty. In Sect. 10.2.4, we conclude with a brief summary of industrial applications of digital twins and simulation input uncertainty.

10.2.1 Understanding Digital Twins

The concept of digital twin was introduced in 2002 by Michael Grieves and later proliferated by Grieves and Vickers (2017). Since then, a significant number of papers have been published on various aspects of digital twins. While it is beyond the scope of this chapter to provide a comprehensive review of the digital twin papers in the literature, we point out a couple of relevant papers that give a general introduction to digital twin development. Among the survey papers that introduce digital twins, digital twin-enabling technologies, applications, case studies, and use cases are Barricelli et al. (2019), Pires et al. (2019), Minerva et al. (2020), Löcklin et al. (2020), Rasheed et al. (2020), Fuller et al. (2020), Mihai et al. (2022), Thelen et al. (2022), and Matta and Lugaresi (2023).

Digital twins have been applied in several sectors including construction (Opoku et al., 2021), (smart) manufacturing (Shao et al., 2019; Overbeck et al., 2021; Kritzinger et al., 2018; Wang et al., 2019), production logistics (Flores-García et al., 2021), agriculture (Verdouw et al., 2021; Kampker et al., 2019), transportation (Ambra & Macharis, 2020), logistics (van der Valk et al., 2022), healthcare (Harper & Mustafee, 2019), and supply chains (Ivanov, 2024). We refer the reader to Liu et al. (2023) and Attaran and Celik (2023) for a comprehensive list of application areas of digital twins. The adaptation of digital twins in industry has been discussed in Kulkarni et al. (2019). More recently, Biller et al. (2023) provided a digital twin development framework for industry practitioners.

10.2.2 Focusing on Digital Twin Simulations

In the simulation communities, dating back to the 1990s, several terms have been introduced to refer to a "digital twin." These terms include *online simulation*, *symbiotic simulation*, and *real-time simulation* (Onggo, 2019). However, these terms have all been dominated by *digital twin* over the last few years. For example, in 2023, the terms real-time simulation and symbiotic simulation appeared only once in the papers in the Proceedings of the Winter Simulation Conference, the flagship conference for simulation practitioners, while the term digital twin appeared more than 30 times. There has been a growing focus on digital twins by the simulation community mainly because simulation is one of the core enabling technologies in digital twin development. There have been panels at the Winter Simulation conference on how digital twins relate to modeling and simulation (Taylor et al., 2021), how they can be enhanced with advances in simulation and artificial intelligence (Taylor et al., 2023), and what a digital twin is for manufacturing research (Akcay et al., 2023). Focusing on process digital twins, Biller et al. (2022) introduce a digital twin framework for simulation practitioners and discuss simulation research areas that are relevant to digital twin development, while Biller et al. (2023) discuss the role of simulation in industrial digital twin development.

Biller and Biller (2023) focus on the role of simulation, machine learning, and reinforcement learning to optimize manufacturing supply chains. The overarching conclusion in these papers is that a (offline) simulation is not enough to qualify for a digital twin. This point has been made in several other papers, including Wooley et al. (2023) and Ali et al. (2024).

10.2.3 Quantifying Simulation Input Uncertainty

A key component of a simulation study is input modeling. It involves selecting probability distributions to represent random inputs to the simulation. Examples of simulation inputs include processing times of workpieces in factory simulations and product demands in supply chain simulations. We use probability distributions to capture the uncertainty in these random variables. There are three main types of uncertainties that may arise in factory and supply chain simulations driven by limited historical data, namely *model uncertainty*, *parameter uncertainty*, and *stochastic uncertainty*. In the remainder of this section, we describe each of them and present an overview of the literature that has addressed their representations in stochastic simulations.

First, we recall the factory simulation example and assume that both part inter-arrival times and machine processing times are exponentially distributed, each with a certain mean. In the presence of historical data for modeling random inputs, the typical practice for picking a distribution is to fit the available data to several distributions using statistical software and select the best-fitting distribution (Law, 2024). However, usually, the software provides us with a couple of good distribution choices, which leads to doubt about whether other distributions (such as gamma) would also do a good job in representing either the inter-arrival times or the processing times. This would be the source of *model uncertainty* in input modeling (Raftery et al., 1996). After we pick a distribution to represent the random inputs, we need to estimate the parameters of this distribution. For example, for the exponential distribution, we need to estimate the mean. The typical practice is to use the MLE method to estimate the distribution parameters. However, the MLEs converge to their true values only in the asymptotic case when an abundant amount of data could be used for the estimation. In reality, we need to work with limited amounts of data, and we are unsure if the estimated mean is true. This is the source of *parameter uncertainty* (Raftery et al., 1996). Finally, because stochastic simulation involves the random number generation from the input probability distributions, there is an inherent uncertainty in the simulation output that is due to the use of random numbers for running the simulation. This type of uncertainty is called *stochastic uncertainty* (Helton, 1997). Stochastic uncertainty in the output data can be controlled by simulating the system under study for a sufficiently large number of replications. However, it is the model uncertainty and the parameter uncertainty, the latter of which is referred to as *input uncertainty* in this chapter, that cannot be controlled by simply increasing the number of replications in a simulation study.

The only way to reduce the input uncertainty is by collecting more data, which may not be possible in certain stages of a simulation study. We provide a formal explanation of this situation, together with a numerical example, in Sect. 10.3.

Several papers in the literature have made an effort to illustrate the negative effects of ignoring input uncertainty in a simulation study. Early work includes Barton (2012) for a queuing simulation, Akcay and Corlu (2017) for inventory simulations, and Corlu et al. (2020) for both. For example, focusing on a simple M/M/1 queuing simulation, Corlu et al. (2020) illustrate that the variability around the mean waiting time highly depends on the estimates of the inter-arrival-time and service-time parameters. The mean waiting time converges to its true value as more data are collected, reducing the impact of the input uncertainty. However, the authors further show that if the input uncertainty is not accounted for properly, increasing the number of replications (hence decreasing the stochastic uncertainty) leads to poor confidence intervals that do not include the true performance measure. These confidence intervals are misleading for decision-makers. While most of the literature focuses on simplistic systems such as the aforementioned M/M/1 queuing system, Biller et al. (2019) focus on a much more complex setting (semiconductor manufacturing) with approximately 700 input random variables. Section 10.4 describes how the authors have chosen to quantify the input uncertainty and how the resulting quantification affected the decision-making process.

Due to the demonstrated adverse effects of ignoring input uncertainty in a simulation study, accounting for input uncertainty has gained increasing attention in the simulation community. There exists a well-established body of work that can be used to account for input uncertainty in a simulation study. The proposed methods include direct bootstrap (Barton & Schruben, 1993, 2001), Bayesian Model Averaging (Chick, 1997, 1999, 2001), (Zouaoui & Wilson, 2003), delta method (Cheng, 1994; Cheng & Holland, 1997, 2004), metamodel-assisted bootstrap (Barton et al., 2010, 2014, 2018), and distributionally robust optimization-based approach (Lam, 2016; Lam & Qian, 2017; Ghosh & Lam, 2019). We refer the reader to Corlu et al. (2020) and Barton et al. (2022) for a comprehensive review of input uncertainty representation and quantification methods.

10.2.4 Industrial Applications of Digital Twins and Input Uncertainty

For examples of industrial digital twin applications, we refer the reader to Reed (2023) for the use of asset twin technology to inform line operators in real time about the product quality at USG, Bakalar et al. (2023) for the optimization of supply chain with advanced analytics and IoT at Georgia-Pacific to improve equipment efficiency, Siemens (2020) for building factory twins using a combination of plant simulation and IoT capabilities to improve production efficiency and quality, Isbill (2023) for reduction in both data cleanup and ingestion time and asset downtime at Lockheed Martin, and Pardue (2023) for the integrated use of IoT data and AI at Volvo Trucks and Mack Trucks.

To the best of our knowledge, Masciarelli and Matta (2022) is the only paper that studies the impact of input uncertainty in digital twins. The authors state that "...it is sometimes impossible, or very costly, to collect sufficient data from which to properly estimate the values of input parameters. This problem is emphasized in digital twin applications, where the simulation is coupled in real time with a system that may change in its behaviour and design and parameters estimation must be repeated at every major change of the real system." Using the metamodel-assisted bootstrapping technique in Barton et al. (2014), the authors show that, even in the presence of limited input data, it is possible to correctly estimate the confidence intervals for the production rate of a lab-scale manufacturing system and to further identify relevant parameters that have the most impact on the production rate. Our goal in this chapter is to illustrate how input uncertainty arises in different phases of digital twin development and discuss how the effect of input uncertainty can be factored into the creation of digital twins for supply chains.

10.3 Problem Description

In this chapter, we consider a factory digital twin built on a dynamic and stochastic simulation of factory operations. We discuss a scenario under which the factory twin is used to support strategic investment decisioning. We also address the factory KPI estimation in the presence of limited data about the history of factory operations. Specifically, we account for inter-arrival-time and processing-time uncertainty and quantify the impact of having limited knowledge about their distribution characterizations on KPIs such as utilization, lead time, inventory, and throughput. Finally, we switch our focus to inventory control and describe the quantification of input uncertainty in the simulation of selected inventory control policies but in the presence of limited demand data. In each of these cases, we assume the use of a discrete-event stochastic simulation to support the management of factory operations.

In this section, we denote a random variable with X and assume that X is exponentially distributed with a mean value of β. However, the (true) value of β is unknown to us. We also assume having access only to a limited history of this random variable, represented by x_n, $n = 1, 2, \ldots, N$. We follow the approach of obtaining an estimate $\hat{\beta}$ of β with the MLE method. Thus, we calculate $\hat{\beta}$ as the mean of the historical dataset (i.e., an unbiased maximum likelihood estimator of β, attaining the minimum variance (Rohatgi & Saleh, 2001)):

$$\hat{\beta} = \bar{x}_N = \frac{1}{N}\sum_{n=1}^{N} x_n.$$

What is important to recognize here is that $\hat{\beta}$ is computed for a single realization of the history. Thus, the value of $\hat{\beta}$ cannot be treated as a constant. In the absence

Table 10.1 Standard deviation and coefficient of variation of $\bar{x}_N$ as a function of $N \in \{10, 30, 50, 100\}$ for the true processing-time parameter $\beta = 0.5$

Statistic	$N = 10$	$N = 30$	$N = 50$	$N = 100$
Standard deviation	0.158	0.091	0.071	0.050
Coefficient of variation	0.316	0.183	0.141	0.100

of prior information about the distribution parameter itself, a way of capturing the uncertainty around $\hat{\beta}$ is through the sampling distribution of MLE of mean processing-time parameter β. The sampling density function tells us which values $\hat{\beta}$ can take and how likely it is for $\hat{\beta}$ to assume those values. Because the random variables X_n, $n = 1, 2, \ldots, N$, are independent and identically distributed, each with an exponential distribution having a true mean parameter of β, the sampling distribution of $\bar{x}_N$ is of type gamma with a shape parameter of N and a scale parameter of β/N. Table 10.1 presents the standard deviation and coefficient of variation of $\hat{\beta}$ for several different values of N when $\beta = 0.5$. It demonstrates the speed at which the sampling distribution of $\bar{x}_N$ approaches the true parameter value β with the increasing length of the history. Section 10.4 discusses how this observation extends to the distributions of the steady-state mean performance measures of single-stage and multistage factory simulations.

If the focus were on a demand random variable X whose true distribution were normal with mean μ and standard deviation σ, then the use of the MLE method—in the presence of only a limited amount of historical data—would lead to the calculation of the sample mean $\bar{x}$ for $\hat{\mu}$ and sample standard deviation s, adjusted for unbiasedness, for $\hat{\sigma}$. It is well known that $\bar{x}$ and s are independent of each other. Furthermore, $\bar{x}$ is normally distributed with mean μ and standard deviation $\sigma/\sqrt{N}$, while s^2 is Gamma distributed with shape parameter $(N-1)/2$ and scale parameter $2\sigma^2/(N-1)$. As discussed in Sect. 10.2, several methods exist to propagate these input uncertainty characterizations through simulated systems. One of these approaches to account for input uncertainty in simulations is to use a simulation replication algorithm. It allows simulation practitioners to capture input uncertainty by sampling input distribution parameters from some probability density functions before each simulation replication. Because it is easy to understand while being insightful, we choose to use this algorithm to describe the problems of propagating input uncertainty in stochastic simulations and quantifying the amounts of stochastic uncertainty and input uncertainty in the simulation output data. Next, we define $\mathbf{x}$ as the vector of historical input data and present a generic description of the simulation replication algorithm to represent the effects of both stochastic uncertainty and input uncertainty in the simulation outputs:

for $m = 1, 2, \ldots, M$ do

  sample the input distribution parameter vector $\boldsymbol{\beta}_m$
  for $r = 1, 2, \ldots, R$ do

    sample the random-number input u_{rm}

perform the simulation run r by using u_{rm} and $\boldsymbol{\beta}_m$
compute the output response $Y_{rm} = \mathbb{H}(u_{rm}, \boldsymbol{\beta}_m)$

end for
compute $Y_{\bullet m}$ as the average of $Y_{rm}, r = 1, 2, \ldots, R$

end for
compute mean $Y_{\bullet\bullet}$ for $\mathbb{E}[Y|\mathbf{x}]$

The simulation replication algorithm presented above includes two nested for-loops, one for representing the input uncertainty and the other for capturing the stochastic uncertainty. More specifically, the outer loop of M replications captures the input uncertainty by sampling a different value for the input distribution parameters of interest, while the inner loop of R replications captures the stochastic uncertainty by sampling from the underlying input distribution function. If we were simulating a system where X introduced earlier in this section is the only random variable of interest, then the parameter vector would have only one component, and sampling input distribution parameter β_m would reduce to sampling a value $\hat{\beta}_m$ from the Gamma distribution with shape parameter N and scale parameter β/N. Performing the simulation run r, on the other hand, corresponds to sampling a realization x from the exponential distribution with parameter $\hat{\beta}_m$. An infinite computing budget is available in each industrial application discussed in Sect. 10.4. Therefore, the simulation replication algorithm is implemented by choosing sufficiently large values for M and R. However, the density function of the input parameter vector might not be as simple as a Gamma distribution function; instead, the input parameter vector might be high dimensional, while its density function might not even have a closed-form expression. Despite sounding more complex than described here, the solution to implementing the simulation replication algorithm in such cases is well documented and utilized by some of the applications discussed in the following section.

Because of the nested for-loops allocated to capture the stochastic uncertainty and the input uncertainty separately, the simulation replication algorithm is often considered an uncertainty decomposition algorithm. More specifically, it writes the output response variance in terms of the stochastic uncertainty and the input uncertainty. Following the presentation in Zouaoui and Wilson (2004), we assume a hierarchical normal model and express the output response from the simulation run r as $Y_{rm} = \mathbb{H}(u_{rm}, \boldsymbol{\beta}_m) = \eta(\boldsymbol{\beta}_m) + \sigma Z_{rm} = \gamma + \psi(\boldsymbol{\beta}_m) + \sigma Z_{rm}$, where Z_{rm}, $r = 1, 2, \ldots, R$ are independent and identically distributed normal random variables, each with a mean of zero and a standard deviation of one. Thus, σZ_{rm} represents the deviation of the simulation output from the response surface $\eta(\boldsymbol{\beta}_m)$ due to the stochastic uncertainty whose source is the random-number input u_{rm}. Furthermore, $\gamma = \mathbb{E}_\beta[\eta(\boldsymbol{\beta}_m)|\mathbf{x}]$ is an unbiased estimator of the mean output response, $\mathbb{E}_\beta[\psi(\boldsymbol{\beta}_m)|\mathbf{x}]$ is equal to zero, and $\mathbb{V}_\beta[\psi(\boldsymbol{\beta}_m)|\mathbf{x}] = \tau^2$ stands for the uncertainty associated with the unknown input parameters. Finally, the response surface representation from Rao (1997) decomposes the output response variance into components of stochastic uncertainty (σ^2) and input uncertainty (τ^2):

$\mathbb{V}[Y_{rm}|\mathbf{x}] = \tau^2 + \sigma^2$. Furthermore, the variance of $Y_{\bullet\bullet}$ (i.e., the average of $Y_{\bullet\, m}, m = 1, 2, \ldots, M$, to estimate $\mathbb{E}[Y|\mathbf{x}]$) can be represented as $\mathbb{V}[Y_{\bullet\bullet}] = \tau^2 + \sigma^2/R$ because

$$
\begin{aligned}
\mathbb{V}[Y_{\bullet\bullet}] &= \mathbb{V}_{\boldsymbol{\beta}}\left[\mathbb{E}\left[R^{-1}\sum_{r=1}^{R}\left(\eta(\boldsymbol{\beta}_m) + \sigma Z_{rm}\right)\middle|\boldsymbol{\beta}\right]\right] \\
&\quad + \mathbb{E}_{\boldsymbol{\beta}}\left[\mathbb{V}\left[R^{-1}\sum_{r=1}^{R}\left(\eta(\boldsymbol{\beta}_m) + \sigma Z_{rm}\right)\middle|\boldsymbol{\beta}\right]\right] \\
&= \mathbb{V}_{\boldsymbol{\beta}}\left[R^{-1}\sum_{r=1}^{R}\eta(\boldsymbol{\beta}_m)\right] + \mathbb{E}_{\boldsymbol{\beta}}\left[R^{-1}\mathbb{V}\left[\sum_{r=1}^{R}(\sigma Z_{rm})\right]\right] \\
&= \mathbb{V}_{\boldsymbol{\beta}}\left[\eta(\boldsymbol{\beta}_m)\right] + \mathbb{E}_{\boldsymbol{\beta}}\left[R^{-1}\sigma^2\sum_{r=1}^{R}\mathbb{V}\left[Z_{rm}\right]\right] \\
&= \tau^2 + R^{-1}\sigma^2.
\end{aligned}
$$

The resulting characterization of the variance of the mean output performance measure provides the insight that increasing the number of simulation replications R decreases the amount of stochastic uncertainty in the confidence interval (i.e., $R^{-1}\sigma^2$), but it has no impact on the portion of the confidence interval length due to input uncertainty (i.e., τ^2). The amount of input uncertainty in the variance of the mean performance measure can be reduced by collecting additional input data; however, how quickly this might happen depends on the underlying input process and the simulation process flow itself. Thus, it is important to account for input uncertainty in the design of simulations to support digital twin development in the presence of limited input data. Because digital twin development occurs in an evolving, learning, and improving environment, there is no stop to acquiring new data and information. For this reason, it is important to understand the importance of capturing input uncertainty and know how to account for both stochastic and input uncertainty in designing digital twin simulations.

10.4 Quantification of Zone-of-Uncertainty in Industrial Use Cases

Our industrial use case involves simulation design and analysis for manufacturing operations management at a semiconductor fabrication facility. Introduced by Biller et al. (2019), the use case illustrates a business analytics effort to design and build a manufacturing facility for a new product developed for a competitive market with high (price and demand) risk exposure. The ultimate goal is to establish a

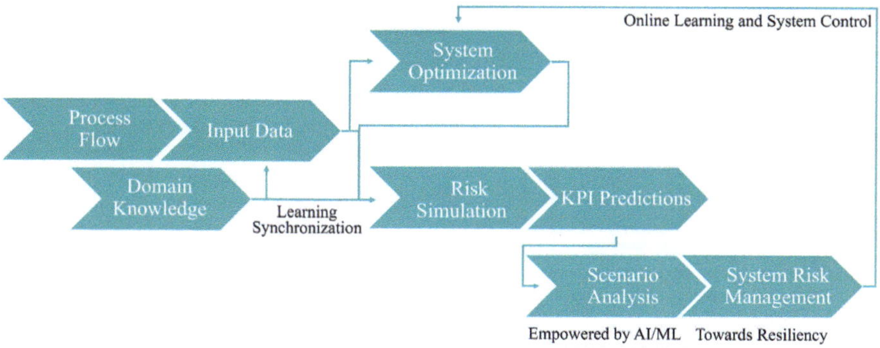

Fig. 10.4 Illustration of a digital twin development environment

production system and manage its activities in such a way as to deliver a unit cost that is low enough to compete successfully in the marketplace over several years. The investment decision timeline starts with strategic equipment portfolio selection, continues with equipment purchase, process verification, and product qualification, and moves toward daily management of operations to reach annual production targets over a horizon of several years. Figure 10.4 illustrates our use case's digital twin development environment. We focus on the digital twin simulation that could support such an industrial use case. We discuss when the concept of zone-of-uncertainty arises and how it can be addressed as part of the digital twin development effort.

In Sect. 10.4.1, we restrict our focus to the selection of an equipment portfolio for a new manufacturing facility. This is the primary strategic investment decision to be made at a time when there does not exist a verified process flow yet. Because no real output data are available, any analytics effort in this strategic phase is to be driven by domain expertise and assumptions reflecting the experts' opinions, the available physical characteristics, and the process limitations provided by the equipment vendors. While historical data might be available from the past production of similar products, this is not the situation in our use case. We expect the development of a discrete-event stochastic simulation to provide insights into what an optimal equipment portfolio would entail. Thus, Sect. 10.4.1 represents the offline model development function of digital twin development.

Section 10.4.2 focuses on what comes after the arrival of the purchased equipment to the new facility. Unpacking each piece of equipment presents an opportunity to collect data to verify the process assumptions made during the offline learning phase described in Sect. 10.4.1. Thus, there arises the opportunity to update simulation input-modeling assumptions and the predictions for the key factory performance indicators with the new data. The ability to perform such updates is critical for the synchronization function of the digital twin development. Figure 10.5 is a high-level illustration of our discussions in Sects. 10.4.1 and 10.4.2.

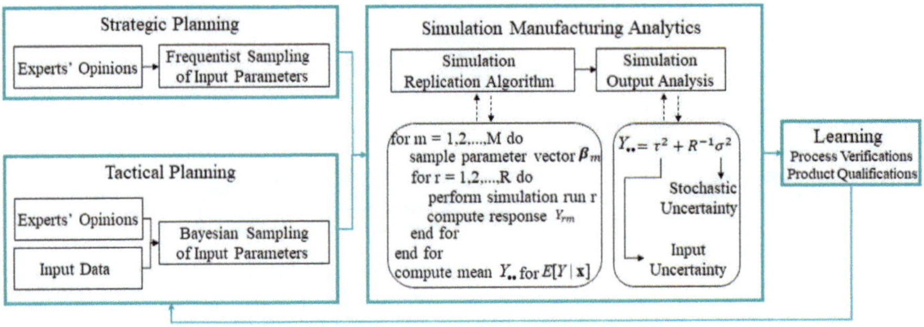

Fig. 10.5 An illustration for the outline of the use case discussion in Sects. 10.4.1 and 10.4.2

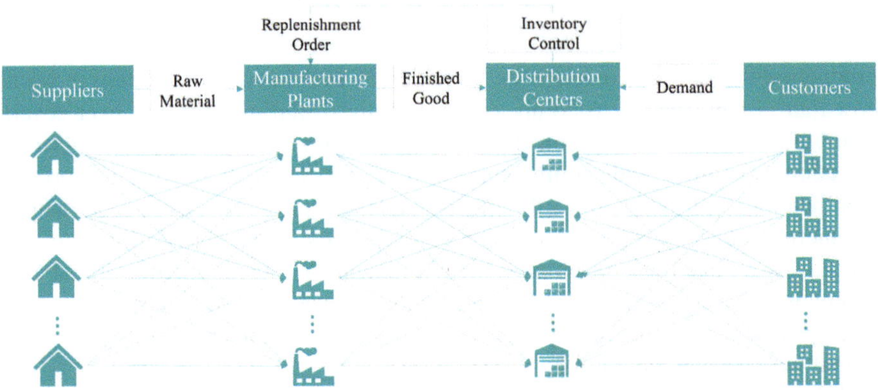

Fig. 10.6 An illustration of a supply chain network

Finally, in Sect. 10.4.3, we switch our focus to inventory control, which is a key issue that arises in the management of supply chains. Figure 10.6 illustrates a supply chain network consisting of all parties involved, directly or indirectly, in fulfilling customers' demands. Distribution centers in this illustration are examples of inventory storage locations. Focusing on any of these locations, Sect. 10.4.3 discusses a framework for setting inventory targets when the demand distribution parameters are unknown and there are only finite amounts of historical demand data to drive the inventory decisions.

10.4.1 Strategic Equipment Portfolio Selection

Biller et al. (2019) develop a discrete-event stochastic simulation to provide decision support for silicon carbide fabrication design and operational policy selection at the New York Power Electronics Manufacturing Consortium facility. This simulation is driven by four different groups of inputs: process flow, recipe, equipment, and shift. Specifically, the process flow contains a number of different steps composed of different recipes, each of which is either a batch process or a cassette-to-cassette process. Recipe description consists of loading and unloading times per cassette as they apply to operators and loading and unloading times per wafer as they apply to fabrication and testing equipment. However, the recipe description also includes the batch and lot sizes, the processing time per batch, minimum and maximum numbers of lots to run, and the number of process chambers for each equipment. Yield and automation level are also considered to relate to a recipe description. Equipment characterization contains time between failures, time to repair, and operator assignments. Finally, shift description contains information about days of production and planned maintenance, shift start times, shift end times, and operator staffing for each shift of a production day. In addition, the use case treats how long it takes to transport a cassette from one step to the next in the process flow as a random variable. As a result, this industrial application is reported to have approximately 700 different input random variables. Sampling realizations from each of these input random variables is an important component of the stochastic simulation development.

Once the discrete-event stochastic simulation capturing the logic of the fabrication process is built, it is executed to obtain predictions of various key performance indicators. Figure 10.7 illustrates that output response is a combination of the simulation's input random variables with the factory's operational policies through the stochastic simulation. First, we assume a set of operating policies (including an inventory policy) and factory configuration (including a factory equipment portfolio). Then, we execute the factory simulation to predict KPIs such as annual throughput (i.e., the total number of chips that are annually placed into packs to be transported from the manufacturing facility to a packaging facility), operator and equipment utilization (i.e., the fractions of time that operator and equipment are busy), inventory (i.e., the total number of lots in the manufacturing facility,

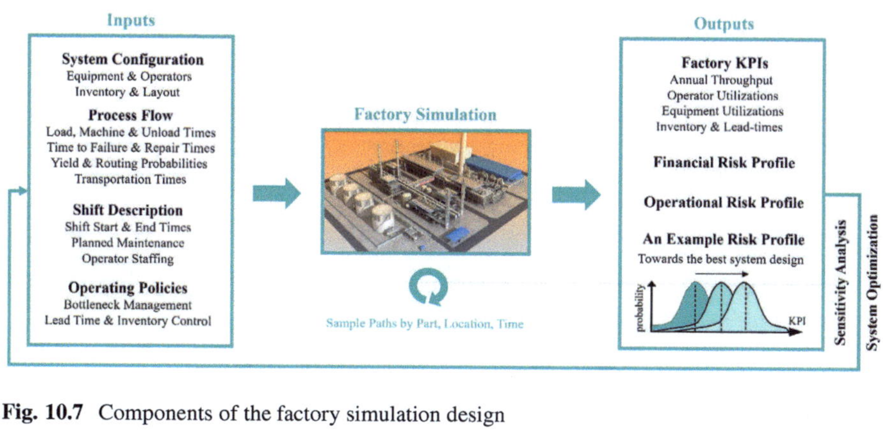

Fig. 10.7 Components of the factory simulation design

including those queueing and those being processed), and lead-time (i.e., the number of calendar days needed to complete the production of a chip). However, the key question is whether a given equipment portfolio would achieve the prespecified annual throughput target. Therefore, the equipment portfolio selection problem can be formulated as a stochastic optimization minimizing the expected shortfall (i.e., the difference between the annual production target and the steady-state estimation of the annual throughput via stochastic simulation) with respect to budget, space, equipment and operator utilization constraints, operator staffing restrictions, and planned maintenance requirements. Thus, simulation plays a key role in identifying annual throughput predictions for various equipment portfolios. For this reason, it is critical to design the stochastic simulation to ensure accurate predictions of the annual throughput. If the input probability distributions of this simulation were known with certainty, then we could control the error in the simulation outputs by choosing the number of simulation replications to produce a standard error that is less than a user-specified percentage of the mean annual throughput and perform steady-state output analysis to remove the initial-condition bias from performance measure calculations (Law, 2024). However, the probability distributions of the input random variables are not known with certainty. The selection of input distribution parameters is to be driven by experts' opinions. A reasonable question to ask here is how the input uncertainty can be quantified and how the resulting quantification might support the strategic equipment portfolio selection.

Biller et al. (2019) model each (continuous) random variable of their factory simulation—except for the times between equipment failures—with a triangular distribution. The triangular distribution is built using three-point estimates elicited from the experts: minimum, mean or mode, and maximum values the corresponding random variable can take. As shown in Fig. 10.8, a triangular distribution can represent symmetric, positively skewed, and negatively skewed distributional shapes. Therefore, it is a suitable candidate to characterize an input random variable when the information is solely elicited from subject matter experts. The time between failures is, on the other hand, modeled by an exponential distribution for each equipment, using the single-point estimates obtained from the available sources of information. Because we can only make assumptions about the distributions of the inputs, we refer to the input uncertainty here as assumption uncertainty and

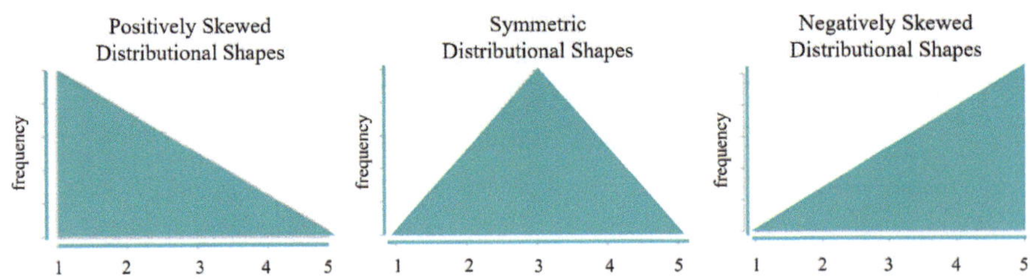

Fig. 10.8 Illustration of the distributional shapes represented by triangular (Beta-PERT) distribution

consider an uncertainty of up to a given $\varphi\%$ around the distribution parameters of each of the simulation inputs in the process flow, recipe, and equipment definitions. Experimenting with a range of values for φ, we execute the simulation replication algorithm, introduced in Sect. 10.3, by sampling the input distribution parameters from the corresponding regions of uncertainty.

Biller et al. (2019) quantify the impact of input uncertainty in mean annual throughput estimation with a method that builds on constructing a $(1-\alpha)100\%$ confidence interval for τ^2/σ^2 (Dean & Voss, 1999) with the simulation output data obtained from the execution of the simulation replication algorithm described in Sect. 10.3:

$$\frac{1}{R}\left[\frac{1}{F_{M-1,M(R-1),1-\alpha/2}}\frac{\text{MST}}{\text{MSE}}-1\right] \le \frac{\tau^2}{\sigma^2} \le \frac{1}{R}\left[\frac{1}{F_{M-1,M(R-1),\alpha/2}}\frac{\text{MST}}{\text{MSE}}-1\right],$$

where $F_{v_1,v_2,\gamma}$ is the γ-quantile of the F distribution with (v_1, v_2) degrees of freedom and

$$\text{MSE} = \frac{M}{R-1}\sum_{m=1}^{M}(Y_{rm} - Y_{\bullet\ m})^2 \text{ and } \text{MST} = \frac{R}{M-1}\sum_{m=1}^{M}(Y_{\bullet\ m} - Y_{\bullet\bullet})^2.$$

Furthermore,

$$\sqrt{\left(\frac{M(R-1)-2}{M(R-1)}\right)\frac{\text{MST}}{\text{MSE}}-1}$$

is an unbiased estimator for $\sqrt{R}\tau/\sigma$.

First, Biller et al. (2019) let φ take the value of 50 and construct a 95% confidence interval for the mean annual throughput by ignoring the input uncertainty. Using the input uncertainty quantification method described in this section, they find that ignoring the input uncertainty overestimates the mean annual throughput by 10%. Furthermore, [15, 129] is identified as the 95% confidence interval for the ratio of the standard deviation due to input uncertainty to the standard deviation due to stochastic uncertainty, while 35 is reported as an unbiased estimate for the expected value of this ratio under no budget constraints on the selection of values for M and R. Then, the management team is asked how much deviation can be tolerated when the confidence interval is reconstructed by accounting for both stochastic and parameter uncertainty in the mean annual throughput estimation. The answer received from expert engineers is a change of $\pm$ 10%. As a result of executing the simulation replication algorithm for different values of φ, it is found that up to a 50% change in distribution parameters would underestimate mean annual throughput by no more than 10% with a probability of 95%. Such a statistical finding often gives management teams confidence in the results obtained from digital twin simulations. Thus, it is of critical importance for digital twin developers to consider uncertainty quantification as discussed in this chapter. In fact, Biller et al. (2019) reported that this statistical finding gave the manufacturing experts confidence in the results obtained and motivated the leadership team to better utilize simulation to gain new insights about the dynamic operational setting even under high uncertainty at the strategic investment stage.

10.4.2 Factory Key Performance Indicator Prediction

As we move from strategic to tactical planning of the factory operations and start collecting data from the equipment installed on the facility floor, we find ourselves in the situation of having new but limited data to update the input distributions that drive the stochastic simulation. Thus, the source of input uncertainty changes from assumptions to the availability of limited historical data collected during process verification and product qualification. This is when the triangular distribution assumption made in the previous section might become a constraint. In this case, it becomes important to work with a family of distribution functions that is able to capture a wide variety of distributional characteristics. Biller et al. (2017) propose the use of a flexible family of distributions to go beyond the use of a triangular distribution to represent the simulation inputs. Specifically, they build on four different transformations of the standard normal distribution function and represent the factory simulation inputs with the Johnson Translation System as they capture any feasible combination of the first four moments—mean, variance, coefficient of skewness, and coefficient of kurtosis—for each simulation input.

In addition to the selection of an input model, another issue is how the input uncertainty can be modeled by using both prior information and new data and how the outer-loop sampling step of the simulation replication algorithm can be conducted. These questions are answered by developing a Bayesian input model. Simply, we select a prior density function that quantifies the prior information about the parameters of the distributions of interest and update the prior density function with the available input data to obtain the posterior density function. We use this posterior density function to sample the input parameter values in the simulation replication algorithm. A posterior density function may have a closed-form expression, making it easy to execute the simulation replication algorithm. For example, for an exponentially distributed random variable X with scale parameter β having the probability density function of the form, $f(X, \beta) = \beta^{-1}\exp(-X\beta^{-1})$, the likelihood function for a historical data of length N, $x_n, n = 1, 2, \ldots, N$, takes the form $\beta^{-N}\exp(-\beta^{-1}\sum_{n=1}^{N} x_n)$. We assume a conjugate inverse gamma prior density function on the parameter β with shape parameter α and scale parameter θ: $\pi(\beta) = \theta^{\alpha}\Gamma^{-1}(\alpha)\beta^{-(\alpha+1)}\exp(-\theta\beta^{-1})$. The conjugate prior ensures that the resulting posterior will be of the same form as the prior with updated parameters. Combining the prior density function with the likelihood function using the Bayes rule (Bernardo & Smith, 2009), we obtain a posterior density function of the form $\beta^{(-N+\alpha+1)}\exp(-(\theta + \sum_{n=1}^{N} x_n)/\beta)$. It is straightforward to sample from this posterior density function as the task reduces to the sampling of β^{-1} from a gamma distribution with shape parameter $N + \alpha$ and scale parameter $(\theta + \sum_{n=1}^{N} x_n)^{-1}$.

However, it may not be straightforward for most distributions, such as those in the Johnson translation system, to sample from the posterior density function due to its functional form. For example, Biller et al. (2017) derive the joint posterior density function for the Johnson log-normal distribution parameters. However, due to the resulting functional form, they resort to an MCMC (Markov Chain Monte Carlo) method for sampling distribution parameter values within the simulation replication algorithm. The idea behind the MCMC method is the simulation of a random walk in the distribution parameter space that converges to the joint posterior density function of the parameters. Then, the parameters sampled in the simulation are averaged to obtain a final estimate of each parameter. Specifically, Biller et al. (2017) discuss the use of the Gibbs sampler, i.e., a widely used MCMC method which requires the sampling of the distribution parameters from their conditional posterior density functions.

Next, we continue to choose annual throughput as the KPI of interest and follow the uncertainty quantification methods described in Sect. 10.4.1 and in this section. First, we construct the 95% confidence interval for the mean annual throughput by accounting for both stochastic and input uncertainty. Then, we compare the ratio of half-length of a 95% confidence interval to the true mean for $R = 10{,}000$ (representing infinite computing budget) and for varying levels of the input data length N. Biller et al. (2017) report the corresponding ratio in percentage as 87% for $N = 10$, 64% for $N = 30$, 41% for $N = 50$, 20% for $N = 100$, and 4% for $N = 500$. Increasing the length of the input data also decreases the difference between the mean annual throughput estimate and its true value, the latter of which is assumed to be known as part of the experimental design. Specifically, the percent difference between the estimated mean and true mean is 10% for $N = 10$, 8% for $N = 30$, 5% for $N = 50$, 3% for $N = 100$, and 0% for $N = 500$. Thus, the simulation output response variance is significantly reduced by additional learning and/or data collected for the input processes of the digital twin simulation. This result is presented by Biller et al. (2017) for the illustrative process flow in Fig. 10.9, and the impact of learning from process verification and product qualification is demonstrated in Fig. 10.10.

Despite the focus on annual throughput estimation, there are other KPIs that inform about the factory's performance. For example, Biller and Mokashi (2018) quantify the impact of limited input data on steady-state manufacturing line perfor-

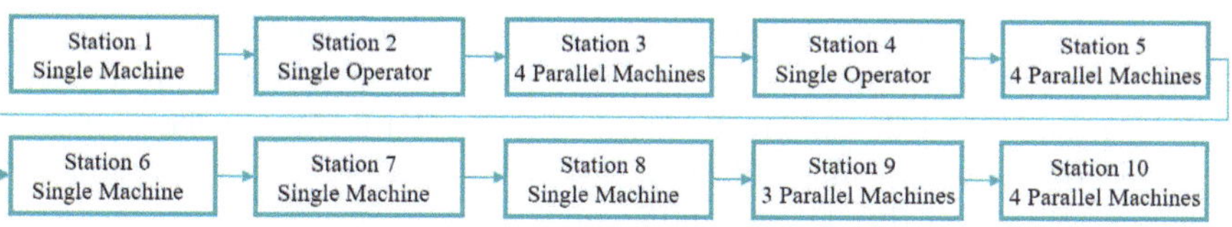

Fig. 10.9 An illustrative example of a factory process flow

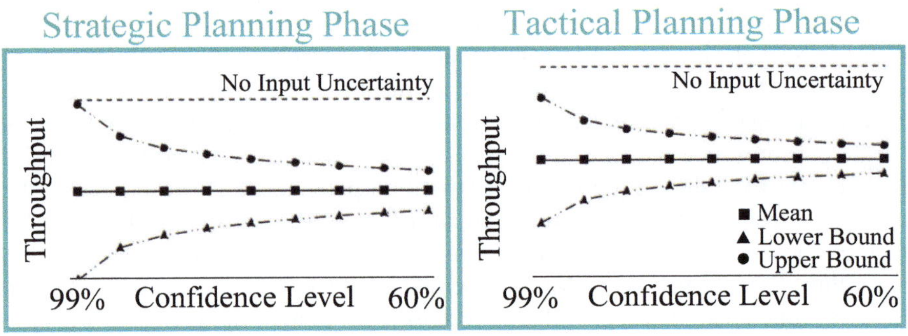

Fig. 10.10 Demonstration of the impact of new data and/or information on mean throughput estimation

mance measures including utilization, lead-time, and inventory. First, experiments are conducted for single-stage manufacturing systems. The simulation practitioner is assumed not to know the true values of the arrival rate and the mean processing time. These unknown parameters are estimated from N inter-arrival-time data points and N service-time data points. An equal length of history for arrival and service processes is chosen to ensure a smaller scale experimental design. The impact of yield loss uncertainty on annual throughput predictions is also studied in the single-stage system. For each input process, it is observed that the input uncertainty due to limited data decreases slowly, especially when a workstation is unreliable with positive yield loss. For example, if N were 10, the coefficient of variation would be 0.32 for the mean service-time parameter and 0.14 for $N = 50$. The coefficients of variations would, however, equal 0.95 and 0.42 for the yield loss parameter, respectively, when $N = 10$ and $N = 50$. The impact of increasing data length on the mean utilization and throughput and their 95% confidence intervals is quantified in Fig. 10.11.

Before the construction of the confidence intervals, it is important to understand whether the density functions of the factory KPIs are symmetric or not. It is noted that the density functions of machine utilization, lead-time, inventory, and throughput are not necessarily symmetric. Therefore, a 95% confidence interval for each of these KPIs is constructed by using 2.5% quantile, 50% quantile, and 97.5% quantile of the simulation output. Also, caution must be exercised in making investment decisions for inventory storage. The 95% confidence interval for average system inventory is between 0.32 parts and 49.13 parts with a median of 1.62 parts when there are only ten observations of each of inter-arrival time and service time. When there are 100 observations in each of the datasets, we identify the upper bound of the 95% confidence interval as ten parts. Therefore, it is important to support buffer decisioning with sufficiently large datasets. Furthermore, the severity of input uncertainty on machine utilization, average inventory, and annual throughput increases with higher (true) system utilization.

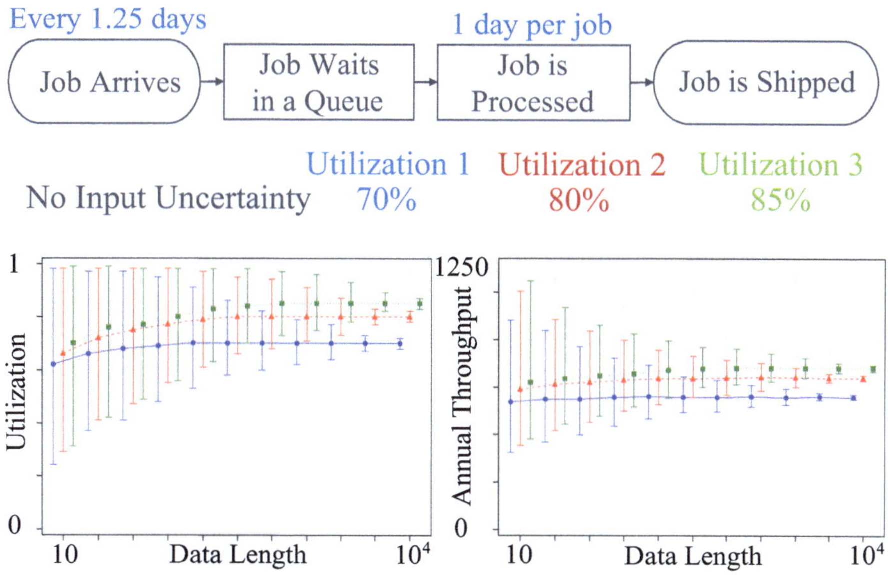

Fig. 10.11 Sensitivity of mean utilization and annual throughput to historical data length

After studying the KPIs for a single-stage system, Biller and Mokashi (2018) switch their focus to multistage manufacturing systems with up to ten serial workstations. Their goal is to enhance understanding of how much uncertainty the KPI predictions are exposed to due to the lack of full knowledge of the systems' arrival and service processes. The length of the service-time data is N for each of the ten workstations on the manufacturing line. Additionally, there are two important issues that arise in a multistage manufacturing system: (i) the impact of input uncertainty on the identification of factory bottlenecks and (ii) the determination of the manufacturing system stages that contribute most to the variability in lead-time and inventory predictions. The authors present a numerical example where it is inconclusive which set of machines is the factory bottleneck. In such a case, it would be beneficial to design a data collection plan in a way to reveal the factory bottlenecks. The decomposition of the lead-time and inventory variability into different manufacturing stages would also be highly beneficial for guiding the collection of new data and information in an effective manner.

10.4.3 Management of Inventory with Limited Demand Data

Most inventory management literature relies on the assumption that input distributions and their parameters are known with certainty. However, this may not be the case in practice, and demand parameter uncertainty arises in addition to stochastic demand uncertainty in digital twin simulation supporting inventory management decisions. Corlu et al. (2018) simulate a newsvendor problem to estimate the mean service level when the demand distribution parameters in the simulation are unknown and estimated from (limited) historical demand data. Using normal and log-normal demand distributions to estimate the demand probability distribution, the authors estimate the parameters of the normal and log-normal distributions assuming the availability of historical demand data of length $N \in \{10, 15, 30, 50, 100, 500, 1000, 10{,}000\}$. Then, they quantify the impact of demand parameter uncertainty on the Type-1 service level (the proportion of time that the demand is satisfied by the on-hand stock) and the Type-2 service level (the proportion of demand satisfied on time from the on-hand stock) using the 95% confidence interval (CI) around the mean service levels. Building on the characterization of the variance of the mean output performance measure $\mathbb{V}[Y_{\bullet\bullet}] = \tau^2 + \sigma^2/R$, the 95% CI around the mean service levels is characterized as $\mathbb{E}[Y_{\bullet\bullet}] \pm 1.96\sqrt{\tau^2 + \sigma^2/R}$. The authors show that the length of the CI can be quite large even when the stochastic uncertainty is negligible. For example, for an estimated Type-1 service level of 0.89, the CI is given by [0.79, 1] which is too wide for an inventory manager to be able to make an informative decision. Another important finding of this study is that while the impact of the demand parameter uncertainty decreases with the increasing length of historical demand data, the decrease is relatively slow, demonstrating the importance of representing the demand parameter uncertainty in inventory system simulations even in the presence of abundant data. For example, when the service level is measured by the Type-1 service level, a 15.8% decrease in the CI length can be achieved with additional five data points, while we would need 50 data points to be able to achieve a 48.82% reduction in the CI length. This means that we would need to increase the number of demand observations at least five times to be able to decrease the CI length by half. For the Type-2 service level, we would even need more data (increase the number of demand observations at least ten times) to be able to decrease the CI length by half.

However, at first glance, how to manage inventory under demand parameter uncertainty may not be obvious. This is the case because we often solve the problem of estimating an inventory target from historical demand data in three steps. We first estimate the unknown demand distribution parameters from the historical data using a standard input-modeling software package. Then, we replace the unknown demand parameters in the expected cost function with their estimates. Finally, we minimize the resulting cost function for obtaining an estimate of the inventory target. For the given demand parameter estimates, this inventory target minimizes the expected cost due to the uncertainty in the demand process. However, this sequential process of estimating the demand parameters first and then solving the inventory optimization problem based on these estimates ignores the demand parameter uncertainty described in Sect. 10.3 and therefore the uncertainty around the inventory-target estimate. Figure 10.12 quantifies the percent contribution of the demand parameter uncertainty to the inaccuracy in cost estimation for a log-normally distributed stochastic process where the value of N changes between 8 and 50 and the scenario of high/medium/low shortage cost is represented by a ratio

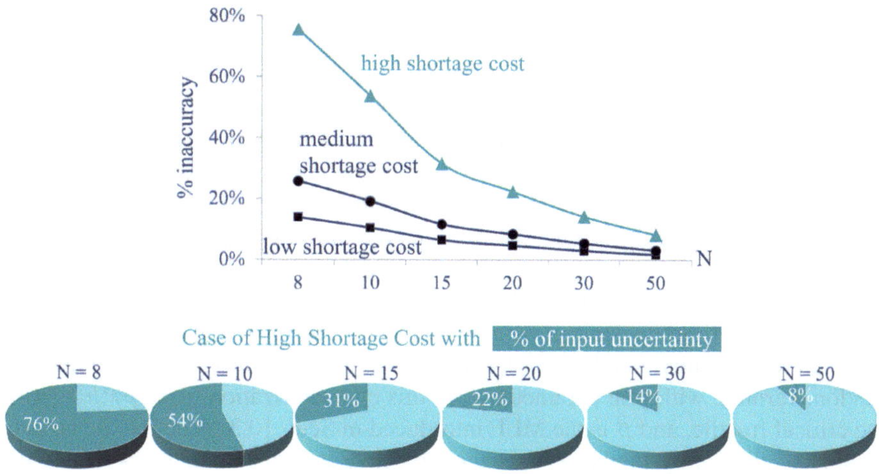

Fig. 10.12 Percent contribution of the demand parameter uncertainty to the overall inaccuracy

of unit shortage cost to unit holding cost that is 99/19/9 (Akcay et al., 2011). A better approach would be combining these three sequential steps into a single one by jointly capturing demand's stochastic and parameter uncertainty and optimizing the expected cost function associated with the inventory policy on hand.

There are several studies that have addressed this challenge. For example, Akcay et al. (2011) present a framework for data-driven inventory management and consider an independent and identically distributed demand process to quantify the inaccuracy in the inventory-target estimation, which is due to demand parameter uncertainty. The key idea behind the development of this framework is to find a balance between the cost of investing in inventory and the risk of making customers unhappy as a result of not carrying sufficient inventory. Actually, the risk-cost trade-off naturally arises in the management of any supply chain. More specifically, Akcay et al. (2011) represent the expected cost as the sum of two distinct components—expected cost due to demand uncertainty and expected cost due to the demand parameter uncertainty—by building on the concept of Expected Total Operating Cost, introduced by Hayes (1969) for inaccuracy quantification in inventory control. The characterization of the first component results in a constant and the expected cost minimization reduces to the minimization of the expected cost due to demand parameter uncertainty. As a function of the amount of the historical demand data, the critical fractile, and the shape parameters of the demand distribution, it is found that the numerical experiments reveal up to 54% inaccuracy in inventory-target estimation for a repeated newsvendor setting, which is a single-product, lost-sales, and periodic-review inventory system. Accounting for demand parameter uncertainty in inventory optimization, as described here, is shown to eliminate up to 37% of the inaccuracy, demonstrating the importance of accounting for demand parameter uncertainty in inventory-target estimation.

Next, we provide an intuitive explanation for the effect of demand parameter uncertainty on the optimal inventory policy. Under complete knowledge of the demand distribution function, it is well known that an increase in shortage (holding) cost would require holding more (less) inventory. It is also the case that newsvendor's loss function $\mathbb{L}(I, x)$ is asymmetric: $h(I - x)$ for $I \geq x$ and $p(x - I)$ for $I < x$, where I is the inventory target and x is the demand realization, while h is the unit inventory holding cost (unit cost of excess stock) and p is the unit cost of shortage. Thus, when the critical fractile (i.e., the ratio of p to the sum of h and p) exceeds 0.5, a unit of unmet demand costs more than a unit held in inventory, and underestimating the safety factor becomes more costly than overestimating it. In the case of ignoring the demand parameter uncertainty, the inventory target is often set to its MLE, which is $\hat{\beta} \log(1/(1-\varphi))$ for an exponentially distributed demand random variable with true parameter β. In this characterization, $\varphi = p/(h+p)$ is the critical fractile, and $\hat{\beta}$ is the MLE introduced in Sect. 10.3. A deliberate upward biasing of the safety factor reduces the expected cost, resulting in the following optimal inventory-target characterization for an exponentially distributed demand process in the presence of only a historical dataset of length N:

$$N\left(\left(\frac{1}{1-\varphi}\right)^{1/(N+1)} - 1\right)\hat{\beta}\log\left(\frac{1}{1-\varphi}\right)$$

It quickly becomes clear that the coefficient in the front of $\hat{\beta}$ serves a positive bias. This means that the demand parameter uncertainty requires holding more inventories when the underlying demand process is exponentially distributed. In the case of a normally distributed demand process, the inventory-target MLE equals $\hat{\mu} + z_\varphi \hat{\sigma}$. Accounting for demand parameter uncertainty, however, leads to the optimal inventory-target characterization with the functional form $\hat{\mu} + \sqrt{(N^2-1)/N^2}\upsilon_{N,\varphi}\hat{\sigma}$, where $\upsilon_{N,\varphi}$ is the Student t distribution's φth quantile with N degrees of freedom. Notice that $\sqrt{(N^2-1)/N^2}\upsilon_{N,\varphi}$ plays the role of a bias factor and provides the following insights: (i) As $N \to \infty$, it holds $(N^2-1)/N^2$ converges to 1 and $\upsilon_{N,\varphi}$ to z_φ, which is the safety factor under full knowledge of the normal distribution function. (ii) If $\sqrt{(N^2-1)/N^2}\upsilon_{N,\varphi} < z_\varphi$ were true, then the optimal action would be holding less inventory, resulting in a negatively biased inventory target. (iii) If $\sqrt{(N^2-1)/N^2}\upsilon_{N,\varphi} > z_\varphi$ were true, then the optimal action would be holding more inventory, resulting in a positively biased inventory target.

In follow-up articles, the assumption of an independent demand process is relaxed, and limited historical auto-correlated demand data and short intermittent demand history are considered. The statistical inventory management practice described here is applied to an industrial case study from a global cosmetics manufacturer struggling with setting inventory targets for its products with different shortage costs and holding costs. The primary challenge is identified as the availability of very limited amounts of historical demand data to drive the inventory decision. The solution of this business case by accounting for parameter uncertainty demonstrates a 23% reduction in the expected cost along with a 2% increase in the expected service level when compared to the approach of ignoring the demand parameter uncertainty. Thus, accounting for input uncertainty, even for a single product, often presents quite promising results.

A more complex multiproduct inventory setting has been considered by Corlu et al. (2017) where the goal is to maximize the joint demand fulfillment probability subject to a budget constraint on the total inventory investment. The authors compare two optimization problems, one of which minimizes the joint demand fulfillment probability obtained with demand parameters estimated from limited historical demand data, and the other minimizes the variance of the demand fulfillment probability, accounting for the uncertainty around the demand parameter estimates. Assuming a P-product inventory setting where each item operates under a periodic-review independent order-up-to policy with deterministic lead-times $L_i, i = 1, 2, \ldots, P$, and a response time window $k \leq L_i, i = 1, 2, \ldots, P$, the joint demand fulfillment probability is defined as the probability of satisfying all item demands within a time window of k periods and takes the following form when item demands are normally distributed with mean parameters $\mu_i, i = 1, 2, \ldots, P$, and standard deviations $\sigma_i, i = 1, 2, \ldots, P$: $y(\mathbf{I}, \Theta) = \prod_{i=1}^{P} \Phi(((I_i - (L_i - k + 1)\mu_i)/(\sqrt{L_i - k + 1}\sigma_i))$. In this characterization, $\mathbf{I} = (I_1, I_2, \ldots, I_P)$ represents the inventory-target vector where I_i is the inventory target for item i and Φ is the standard normal cumulative distribution function. The authors characterize the demand parameter uncertainty around the joint demand fulfillment probability $y(\mathbf{I}, \Theta)$ by building a Bayesian model for the parameters $\Theta = (\theta_1, \theta_2, \ldots, \theta_P)^{'}$, where $\theta_i = (\mu_i, \sigma_i)^{'}$ for $i = 1, 2, \ldots, P$. A normal-inverse gamma density function is used as the prior density function, and the posterior density function is obtained by updating the demand parameters with the historical data. At the heart of the analysis is the fact that the joint distribution of the parameter vector Θ asymptotically converges to a multivariate normal distribution with mean $\tilde{\Theta} = \left(\mu_{i,N_i}, \sqrt{\zeta_{i,N_i}^2/\nu_{i,N_i} + 3}\right)$ and variance-covariance matrix G having the diagonals represented by $G_{i,N_i} = [\zeta_{i,N_i}^2/(\kappa_{i,N_i}(\nu_{i,N_i} + 3))\ 0;\ 0\ 2\zeta_{i,N_i}^2/(\nu_{i,N_i} + 3)^3]$. In this representation, $\kappa_{i,N_i} = \kappa_{i,0} + N_i$, $\nu_{i,N_i} = \nu_{i,0} + N_i$, $\mu_{i,N_i} = (\kappa_{i,0}\mu_{i,0} + \sum_{n=1}^{N_i} d_i, n)/(\kappa_{i,0} + N_i)$, and

$$\zeta_{i,N_i}^2 = \zeta_{i,0}^2 + \frac{\kappa_{i,0} N_i}{\kappa_{i,0} + N_i} \left(\frac{\sum_{n=1}^{N_i} d_{i,n}}{N_i} - \mu_{i,0} \right)^2 + \sum_{n=1}^{N_i} \left(d_{i,n} - \frac{\sum_{n=1}^{N_i} d_{i,n}}{N_i} \right)^2,$$

where $\kappa_{i,0}$, $\zeta_{i,0}^2$, $\mu_{i,0}$, and $\nu_{i,0}$ are prior hyperparameters estimated from expert opinion. What is important to recognize here is that the distribution of $\tilde{\Theta}$ induces a distribution on $y(\mathbf{I}, \Theta)$. Thus, the variance of $y(\mathbf{I}, \Theta)$ that is due to the demand parameter uncertainty is characterized as follows:

$$\mathrm{Var}(\mathrm{y}(\mathbf{I}, \Theta)) = \sum_{i=1}^{P} \left(\frac{L_i - k + 1}{\kappa_{i,N_i}} + \frac{(I_i - (L_i - k + 1)\tilde{\mu}_i)^2 (\nu_{i,N_i} + 3)^{-1}}{2(L_i - k + 1)\tilde{\sigma}_i^2} \right)$$

$$\prod_{j=1, j\neq i}^{P} \Phi^2\left(\frac{I_j-(L_j-k+1)\tilde{\mu}_j}{\sqrt{L_j-k+1}\tilde{\sigma}_j}\right)\phi^2\left(\frac{I_i-(L_i-k+1)\tilde{\mu}_i}{\sqrt{L_i-k+1}\tilde{\sigma}_i}\right),$$

with $\tilde{\mu}_i$ and $\tilde{\sigma}_i$ given by μ_{i,N_i} and $\sqrt{\zeta_{i,N_i}^2/\nu_{i,N_i}+3}$, respectively. Thus, the first optimization problem, which ignores the demand parameter uncertainty, maximizes $y(\mathbf{I}, \Theta)$ by replacing Θ with MLE estimates in the objective function. In contrast, the optimization problem accounting for the demand parameter uncertainty minimizes $\mathrm{Var}(\mathrm{y}(\mathbf{I}, \Theta))$. The performance of these two approaches is compared by the 3σ CI width that they yield. The experiments demonstrate the value of minimizing the variance of the demand fulfillment probability, especially with an increasing number of products. For example, in a two-product setting with ten demand data points for each product, the 3σ CI width is 0.53 when ignoring demand parameter uncertainty and 0.43 when accounting for the demand parameter uncertainty. This is about a 19% decrease in the CI width. For 20 products, the reduction in the CI width is much higher (about 44.8%).

10.5 Conclusion

The goal of this chapter is to provide practitioners guidance about how to develop digital twin simulations in the presence of limited data. We give an overview of the industrial applications where the representation of zone-of-uncertainty arises as a challenge in digital twin development. We not only describe how this challenge has been addressed but also inform the reader about the existing body of knowledge that can be utilized for risk awareness when simulation is chosen as a digital twin-enabling technology.

References

Akcay, A., Biller, B., & Tayur, S. (2011). Improved inventory targets in the presence of limited historical demand data. *Manufacturing & Service Operations Management, 13*(3), 297–309.

Akcay, A., Biller, S., Gan, B.P., Laroque, C., & Shao, G. (2023). Maintenance and Operations of Manufacturing Digital Twins. In *Proceedings of the 2023 Winter Simulation Conference* (pp. 1888–1899). IEEE.

Akcay, A., & Corlu, C.G. (2017). Simulation of inventory systems with unknown input models: a data-driven approach. *International Journal of Production Research, 55*(19), 5826–5840.

Ali, Z., Biglari, R., Denil, J., Mertens, J., Poursoltan, M., & Traoré, M.K. (2024). From modeling and simulation to Digital Twin: evolution or revolution? *Simulation, 100*(7), 751–769.

Ambra, T., & Macharis, C. (2020). Agent-based digital twins (ABM-DT) in synchromodal transport and logistics: The fusion of virtual and physical spaces. In *Proceedings of the 2020 Winter Simulation Conference* (pp. 159–169). IEEE.

Attaran, M., & Celik, B.G. (2023). Digital Twin: Benefits, use cases, challenges, and opportunities. *Decision Analytics Journal, 6*, 100165.

Bakalar, S., Shah, R., & Berryman, T. (2023). Georgia-Pacific relies on SAS®Viya®on Amazon Web Services to improve equipment efficiency, reduce downtime, optimize shipping logistics and predict customer churn. https://www.sas.com/en_sg/customers/georgia-pacific.html, Accessed 10 July 2024.
Barricelli, B.R., Casiraghi, E., & Fogli, D. (2019). A survey on digital twin: Definitions, characteristics, applications, and design implications. *IEEE Access, 7*, 167653–167671.
Barton, R.R. (2012). Tutorial: Input uncertainty in output analysis. In *Proceedings of the 2012 Winter Simulation Conference* (pp. 1–12). IEEE.
Barton, R.R., Lam, H., & Song, E. (2018). Revisiting direct bootstrap resampling for input model uncertainty. In *Proceedings of the 2018 Winter Simulation Conference* (pp. 1635–1645). IEEE.
Barton, R.R., Lam, H., & Song, E. (2022). *Input uncertainty in stochastic simulation* (pp. 573–620). Springer International Publishing.
Barton, R.R., Nelson, B.L., & Xie, W. (2010). A framework for input uncertainty analysis. In *Proceedings of the 2010 Winter Simulation Conference* (pp. 1189–1198). IEEE.
Barton, R.R., Nelson, B.L., & Xie, W. (2014). Quantifying input uncertainty via simulation confidence intervals. *INFORMS Journal on Computing, 26*(1), 74–87.
Barton, R.R., & Schruben, L.W. (1993). Uniform and bootstrap resampling of empirical distributions. In *Proceedings of the 1993 Winter Simulation Conference* (pp. 503–508). IEEE.
Barton, R.R., & Schruben, L.W. (2001). Resampling methods for input modeling. In *Proceedings of the 2001 Winter Simulation Conference (Cat. No. 01CH37304)* (vol. 1, pp. 372–378). IEEE.
Bernardo, J.M., & Smith, A.F.M. (2009). *Bayesian theory* (vol. 405). John Wiley & Sons.
Biller, B., & Biller, S. (2023). Implementing digital twins that learn: AI and simulation are at the core. *Machines, 11*(4), 425.
Biller, B., Biller, S.R., Dulgeroglu, O., & Corlu, C.G. (2017). The role of learning on industrial simulation design and analysis. In *Proceedings of the 2017 Winter Simulation Conference* (pp. 3287–3298). IEEE.
Biller, B., Hartig, M., Olson, R.J., Sandvik, P., Trant, G., Sui, Y., & Minnick, A. (2019). General Electric uses simulation and risk analysis for silicon carbide production system design. *INFORMS Journal on Applied Analytics, 49*(2), 117–128.
Biller, B., Jiang, X., Yi, J., Venditti, P., & Biller, S. (2022). "Simulation: The critical technology in digital twin development. In *Proceedings of the 2022 Winter Simulation Conference* (pp. 1340–1355). IEEE.
Biller, B., & Mokashi, A.C. (2018). Propagation of input uncertainty in manufacturing process flow simulations. In *Proceedings of the 2018 Winter Simulation Conference* (pp. 3144–3155). IEEE.
Biller, B., Yi, J., & Biller, S. (2023). A practitioner's guide to digital twin development. In *Tutorials in Operations Research: Advancing the Frontiers of OR/MS: From Methodologies to Applications* (pp. 198–227). INFORMS.
Biller, S., Venditti, P., Yi, J., Jiang, X., & Biller, B. (2023). Simulation-driven digital twins: The DNA of resilient supply chains. In *Proceedings of the 2023 Winter Simulation Conference* (pp. 1463–1477). IEEE.
Cheng, R.C.H. (1994). Selecting input models. In *Proceedings of the 1994 Winter Simulation Conference* (pp. 184–191). IEEE.
Cheng, R.C.H., & Holland, W. (1997). Sensitivity of computer simulation experiments to errors in input data. *Journal of Statistical Computation and Simulation, 57*(1-4), 219–241.
Cheng, R.C.H., & Holland, W. (2004). Calculation of confidence intervals for simulation output. *ACM Transactions on Modeling and Computer Simulation (TOMACS), 14*(4), 344–362.
Chick, S.E. (1997). Bayesian analysis for simulation input and output. In *Proceedings of the 1997 Winter Simulation Conference* (pp. 253–260). IEEE.
Chick, S.E. (1999). Steps to implement Bayesian input distribution selection. In *Proceedings of the 1999 Winter Simulation Conference: Simulation – a bridge to the future – Volume 1* (pp. 317–324). IEEE.
Chick, S.E. (2001). Input distribution selection for simulation experiments: Accounting for input uncertainty. *Operations Research, 49*(5), 744–758.

Corlu, C.G., Akcay, A., & Xie, W. (2020). Stochastic simulation under input uncertainty: A review. *Operations Research Perspectives, 7*, 100162.

Corlu, C.G., Biller, B., & Tayur, S. (2017). Demand fulfillment probability in a multi-item inventory system with limited historical data. *IISE Transactions, 49*(12), 1087–1100.

Corlu, C.G., Biller, B., & Tayur, S. (2018). Driving inventory system simulations with limited demand data: Insights from the newsvendor problem. *Journal of Simulation, 13*(2), 152–162.

Dean, A., & Voss, D. (1999). *Design and analysis of experiments*. Springer.

Digital-Twin-Tech. (2024). Digital twins market by technology, twinning type, cyber-to-physical solutions, use cases and applications in industry verticals 2024–2029. https://www.businesswire.com/news/home/20240708391066/en/Digital-Twins-Market-Report-2024-2029, Accessed 13 July 2024.

Flores-García, E., Jeong, Y., Wiktorsson, M., Liu, S., Wang, L., & Kim, G. (2021). Digital twin-based services for smart production logistics. In *Proceedings of the 2021 Winter Simulation Conference* (pp. 1–12). IEEE.

Fuller, A., Fan, Z., Day, C., & Barlow, C. (2020). Digital twin: Enabling technologies, challenges and open research. *IEEE Access, 8*, 108952–108971.

Gartner. (2023). Top strategic technology trends 2023. https://emt.gartnerweb.com/ngw/globalassets/en/publications/documents/2023-gartner-top-strategic-technology-trends-ebook.pdf, Accessed 13 July 2024.

Ghosh, S., & Lam, H. (2019). Robust analysis in stochastic simulation: Computation and performance guarantees. *Operations Research, 67*(1), 232–249.

Grieves, M., & Vickers, J. (2017). Digital twin: Mitigating unpredictable, undesirable emergent behavior in complex systems. *Transdisciplinary perspectives on complex systems: New findings and approaches* (pp. 85–113).

Harper, A., & Mustafee, N. (2019). A hybrid modelling approach using forecasting and real-time simulation to prevent emergency department overcrowding. In *Proceedings of the 2019 Winter Simulation Conference* (pp. 1208–1219). IEEE.

Hayes, R.H. (1969). Statistical estimation problems in inventory control. *Management Science, 15*(11), 686–701.

Helton, J.C. (1997). Uncertainty and sensitivity analysis in the presence of stochastic and subjective uncertainty. *Journal of Statistical Computation and Simulation, 57*(1-4), 3–76.

Isbill, M. (2023). Lockheed Martin revolutionizes aircraft maintenance with the SAS platform. https://www.sas.com/en_sa/customers/lockheed-martin.html, Accessed 10 July 2024.

Ivanov, D. (2024). Conceptualisation of a 7-element digital twin framework in supply chain and operations management. *International Journal of Production Research, 62*(6), 2220–2232.

Kampker, A., Stich, V., Jussen, P., Moser, B., & Kuntz, J. (2019). Business models for industrial smart services–the example of a digital twin for a product-service-system for potato harvesting. *Procedia Cirp, 83*, 534–540.

Kritzinger, W., Karner, M., Traar, G., Henjes, J., & Sihn, W. (2018). Digital Twin in manufacturing: A categorical literature review and classification. *Ifac-PapersOnline, 51*(11), 1016–1022.

Kulkarni, V., Barat, S., & Clark, T. (2019). Towards adaptive enterprises using digital twins. In *Proceedings of the 2019 Winter Simulation Conference* (pp. 60–74). IEEE.

Lam, H. (2016). Robust sensitivity analysis for stochastic systems. *Mathematics of Operations Research, 41*(4), 1248–1275.

Lam, H., & Qian, H. (2017). Optimization-based quantification of simulation input uncertainty via empirical likelihood. *arXiv:1707.05917*.

Law, A.M. (2024). *Simulation modeling and analysis*. 6th edn. Mcgraw-Hill.

Liu, X., Jiang, D., Tao, B., Xiang, F., Jiang, G., Sun, Y., Kong, J., & Li, G. (2023). A systematic review of digital twin about physical entities, virtual models, twin data, and applications. *Advanced Engineering Informatics, 55*, 101876.

Löcklin, A., Müller, M., Jung, T., Jazdi, N., White, D., & Weyrich, M. (2020). Digital twin for verification and validation of industrial automation systems–a survey. In *2020 25th IEEE International Conference on Emerging Technologies and Factory Automation (ETFA)* (vol. 1, pp. 851–858). IEEE.

Masciarelli, A., & Matta, A. (2022). Managing input parameter uncertainty in digital twins. *IFAC-PapersOnLine, 55*(2), 13–18.

Matta, A., & Lugaresi, G. (2023). Digital twins: Features, models, and services. In *Proceedings of the 2023 Winter Simulation Conference* (pp. 46–60). IEEE.

McKinsey & Company. (2021). The Internet of Things: Catching up to an accelerating opportunity. https://www.mckinsey.com/~/media/mckinsey/business%20functions/mckinsey%20digital/our%20insights/iot%20value%20set%20to%20accelerate%20through%202030%20where%20and%20how%20to%20capture%20it/the-internet-of-things-catching-up-to-an-accelerating-opportunity-final.pdf, Accessed 13 July 2024.

Mihai, S., Yaqoob, M., Hung, D.V., Davis, W., Towakel, P., Raza, M., Karamanoglu, M., et al. (2022). Digital twins: A survey on enabling technologies, challenges, trends and future prospects. *IEEE Communications Surveys & Tutorials, 24*(4), 2255–2291.

Minerva, R., Lee, G.M., & Crespi, N. (2020). Digital twin in the IoT context: A survey on technical features, scenarios, and architectural models. *Proceedings of the IEEE, 108*(10), 1785–1824.

Onggo, B.S. (2019). Symbiotic simulation system (S3) for industry 4.0. *Simulation for Industry 4.0: Past, Present, and Future,* 153–165.

Opoku, D.-G.J., Perera, S., Osei-Kyei, R., & Rashidi, M. (2021). Digital twin application in the construction industry: A literature review. *Journal of Building Engineering, 40*, 102726.

Overbeck, L., Le Louarn, A., Brützel, O., Stricker, N., Lanza, G., Franke, J., & Schuderer, P. (2021). Continuous validation and updating for high accuracy of digital twins of production systems. *Simulation in Produktion und Logistik*, 609–617.

Pardue, D. (2023). Volvo Trucks and Mack Trucks use sensor data and SAS AI solutions to minimize unplanned downtime. https://www.sas.com/en_th/customers/volvo-trucks-mack-trucks.html, Accessed 10 July 2024.

Pires, F., Cachada, A., Barbosa, J., Moreira, A.P., & Leitão, P. (2019). Digital twin in industry 4.0: Technologies, applications and challenges. In *2019 IEEE 17th International Conference on Industrial Informatics (INDIN)* (vol. 1, pp. 721–726). IEEE.

Raftery, A.E., Madigan, D., & Volinsky, C.T. (1996). Accounting for model uncertainty in survival analysis improves predictive performance. *Bayesian Statistics, 5*, 323–349.

Rao, P.S.R.S. (1997). *Variance components: mixed models, methodologies and applications* (vol. 78). CRC Press.

Rasheed, A., San, O., & Kvamsdal, T. (2020). Digital twin: Values, challenges and enablers from a modeling perspective. *IEEE Access, 8*, 21980–22012.

Reed, P. (2023). USG Corporation relies on SAS to improve its manufacturing process and reduce downtime, costs, and energy consumption. Customer success stories. https://www.sas.com/en_us/customers/usg.html, Accessed 10 July 2024.

Rohatgi, V.K., & Saleh, A.K.M.E. (2001). Nonparametric statistical inference. *An introduction to probability and statistics*, 2nd edn. Wiley.

Shao, G., Jain, S., Laroque, C., Lee, L.H., Lendermann, P., & Rose, O. (2019). Digital twin for smart manufacturing: the simulation aspect. In *Proceedings of the 2019 Winter Simulation Conference* (pp. 2085–2098). IEEE.

Siemens. (2020). Siemens (Global Cloud Practice, Berlin, Germany). Closed Loop Plant Simulation Solution Presentation.

Taylor, S.J.E., Johansson, B., Jeon, S., Lee, L.H., Lendermann, P., & Shao, G. (2021). Using simulation and digital twins to innovate: are we getting smarter? In *Proceedings of the 2021 Winter Simulation Conference* (pp. 1–13). IEEE.

Taylor, S.J.E., Macal, C.M., Matta, A., Rabe, M., Sanchez, S.M., & Shao, G. (2023). Enhancing digital twins with advances in simulation and artificial intelligence: opportunities and challenges. In *Proceedings of the 2023 Winter Simulation Conference* (pp. 3296–3310). IEEE.

Thelen, A., Zhang, X., Fink, O., Lu, Y., Ghosh, S., Youn, B.D., Todd, M.D., Mahadevan, S., Hu, C., & Hu, Z. (2022). A comprehensive review of digital twin—part 1: modeling and twinning enabling technologies. *Structural and Multidisciplinary Optimization, 65*(12), 354.

van der Valk, H., Strobel, G., Winkelmann, S., Hunker, J., & Tomczyk, M. (2022). Supply chains in the era of digital twins–a review. *Procedia Computer Science, 204*, 156–163.

Verdouw, C., Tekinerdogan, B., Beulens, A., & Wolfert, S. (2021). Digital twins in smart farming. *Agricultural Systems, 189*, 103046.
Wang, J., Ye, L., Gao, R.X., Li, C., & Zhang, L. (2019). Digital Twin for rotating machinery fault diagnosis in smart manufacturing. *International Journal of Production Research, 57*(12), 3920–3934.
Wooley, A., Silva, D.F., & Bitencourt, J. (2023). When is a simulation a digital twin? A systematic literature review. *Manufacturing Letters, 35*, 940–951.
Zouaoui, F., & Wilson, J.R. (2003). Accounting for parameter uncertainty in simulation input modeling. *IIE Transactions, 35*(9), 781–792.
Zouaoui, F., & Wilson, J.R. (2004). Accounting for input-model and input-parameter uncertainties in simulation. *IIE Transactions, 36*(11), 1135–1151.

Part IV
Specific Applications and Case Studies

Chapter 11
Digitalizing the Automotive Assembly Supply Chain Using Multi-Agent-Based Digital Twins

Konstantinos Mykoniatis and Michail Katsigiannis

Abstract The automotive industry encounters increasing supply chain challenges due to fluctuating demand, inventory imbalances, and long lead times. Traditional approaches face economic uncertainties, supplier disruptions, and trade policy changes, leading to inefficiencies and cost volatility. To address these issues, industry is adopting Industry 4.0 technologies, particularly Digital Twins, which create real-time virtual representations of physical systems, integrating sensor data and automation for enhanced decision-making, and predictive analytics.

This chapter presents a Multi-Agent-Based Digital Twin (MABDT) model targeted toward automotive manufacturing supply chains. The proposed three-layer architecture, utilizing Physical, Digital Twin, and Application Layers, leverages Agent-Based Modeling simulation techniques to create real-time monitoring agents of supply chain elements using statecharts and structured agent behaviors.

A case study on the Tiger Motors simulated assembly line demonstrates how this architecture is used to model cars, workstations, material supermarkets, and manufacturing cells as agents that interact dynamically with each other to develop applications that can improve takt time management, production visibility, and operational efficiency. Integrating real-time data with Digital Twin agents has the potential to enable predictive maintenance, reduce downtime, and ensure better supply-production synchronization.

By aligning Digital Twins with multi-agent system architectures, this work establishes a scalable framework for intelligent manufacturing systems. Through this work we aim to contribute to adaptive supply chain management, bridging theory, and practice while serving as an educational testbed for digital manufacturing in automotive manufacturing.

Konstantinos Mykoniatis and Michail Katsigiannis are contributed equally to this work.

K. Mykoniatis (✉) · M. Katsigiannis
Department of Industrial and Systems Engineering, Auburn University, Auburn, AL, USA
e-mail: kmykoniatis@auburn.edu; mkatsigiannis@auburn.edu

G. Rabadi, B. Soykan (eds.), *Optimizing Supply Chains Through Digital Twins*,
International Series in Operations Research & Management Science 375,
https://doi.org/10.1007/978-3-032-08147-6_11

Keywords Digital Twin · Automotive manufacturing · Multi-agent simulation · Agent-based modeling · Industry 4.0 · Internet of Things

11.1 Introduction

The automotive industry faces increasing challenges in managing its supply chain efficiently, as fluctuations in demand, inventory imbalances, and long lead times create significant operational complexities (Zavala-Alcívar et al., 2020). Traditional supply chain systems struggle to adapt to shifting consumer preferences, economic uncertainties, and disruptions in global supplier networks. Trade policies and tariffs also contribute to cost volatility and sourcing difficulties (Dong & Kouvelis, 2020). These inefficiencies lead to inconsistent production schedules, excess inventory, and delays in sourcing critical components, affecting cost management and operational performance.

The growing need for digital transformation in supply chain management has become evident in many industries (Tiwari et al., 2024), and the automotive sector is no exception. The industry is undergoing a digital transformation to address these challenges, using Industry 4.0 (I4.0) technologies to enhance real-time monitoring, automation, predictive analytics, and supply chain agility. Among the I4.0 pillars, Digital Twin (DT) technologies have emerged as a powerful tool to improve production efficiency, inventory management, and supply chain visibility (Stroumpoulis & Kopanaki, 2022). A DT is a virtual representation of a system (physical or cyber-physical) that replicates real-time behavior by integrating real-time data from sensors, Internet of Things (IoT) devices, and automation systems. By establishing bidirectional communication between the actual system and its virtual representation using real-time insights, Digital Twins (DTs) enable manufacturers to anticipate disruptions, reduce downtime, and improve decision-making (Mykoniatis & Harris, 2021).

Lean Manufacturing (LM) adoption in the manufacturing industry has been widely studied, with research indicating its positive impact on operational performance in both developed and developing countries (Maware et al., 2022). However, implementation challenges persist, including organizational culture change, management support, and employee resistance (Maware & Parsley, 2022). The integration of LM with Industry 4.0 technologies has shown promising results, enhancing efficiency and productivity (Narula et al., 2023). However, real-world implementation of these principles remains challenging due to the complexities of balancing takt time, cycle time, and work-in-progress (WIP) inventory. DTs offer a promising approach to overcoming these challenges by providing a dynamic and data-driven framework for optimizing production systems in real time (Pantelidakis et al., 2022; Pantelidakis & Mykoniatis, 2024).

This chapter presents a Multi-Agent-Based Digital Twin (MABDT) architecture that can be used to design DTs for automotive manufacturing supply chains. By leveraging multi-agent systems, the MABDT framework aims to replicate and improve assembly line operations through autonomous agents that represent key system elements such as workstations, operators, and automation systems. These agents interact dynamically based on real-time data, enabling adaptive decision-making, scenario testing, and proactive error detection.

To illustrate the practical application of this approach, we present a case study on the Tiger Motors facility, a simulated automotive assembly line used for educational and research purposes. The Tiger Motors facility provides a controlled environment for exploring the integration of digitalization, LM, and DT technology. In this chapter, we describe how real-time data from Tiger Motors' Programmable Logic Controllers (PLCs) and IoT devices are incorporated into the MABDT framework to generate actionable insights.

By integrating a multi-agent-based modeling approach, real-time analytics, and LM principles, this chapter highlights the transformative potential of MABDT in automating decision-making, optimizing production processes, and strengthening supply chain resilience in the modern automotive industry.

11.2 Background

11.2.1 Digital Twins in Supply Chain Management

In the context of automotive assembly lines, DTs create a real-time dynamic model that enables manufacturers to monitor, simulate, predict, and optimize assembly line operations, improving efficiency and reducing downtime. The concept of DTs, introduced by Dr. Michael Grieves in 2002 within product life cycle management, has evolved to encompass entire production systems (Grieves, 2024). Advances in IoT, big data analytics, artificial intelligence, and cloud computing have enabled real-time monitoring and predictive capabilities, making DTs a valuable tool for optimizing assembly line performance. Manufacturers gain visibility into bottlenecks, equipment health, and material flow by digitally replicating their production processes, enabling data-driven decision-making and process improvements. DTs can improve supply chain connectivity, visibility, and agility, supporting intelligent, and integrated operations (Wang et al., 2022). Recent research has focused on developing conceptual frameworks for the implementation of DT in logistics and supply chain systems, addressing challenges, and identifying research opportunities (Le & Fan, 2024). Although DT applications in system design and production planning have increased, their use in supply chain management remains limited (Lugaresi et al., 2023). However, the growing complexity of global supply networks and the need for short-term planning capabilities make DTs a promising solution for effective supply chain management.

Predictive maintenance is another key advantage, as DTs can monitor machinery health and forecast failures before they occur, reducing unplanned downtime and improving overall productivity.

Automotive leaders such as BMW and Ford have already integrated Digital Twin technology to enhance assembly line efficiency (Ford Motor Company, 2023; BMW Group, 2022). By simulating different production scenarios, manufacturers can assess the impact of changes in demand, supply chain disruptions, or process modifications before implementing them in the physical world. This ensures optimal resource utilization, minimizes waste, and improves responsiveness to market fluctuations.

11.2.2 Automotive Assembly Line Supply Chain Challenges

The automotive assembly line faces several core challenges in supply chain management, including demand volatility, forecasting difficulties, long lead times, and complex inventory control (Reuters, 2025; Financial Times, 2025). Assembly lines operate in fast-paced and highly competitive environments where fluctuations in consumer demand, economic conditions, and supply chain disruptions can directly impact production efficiency. Variability in demand often leads to misalignment in parts availability, resulting in costly production stoppages or excess inventory accumulation.

Long lead times further complicate assembly line operations. Automotive manufacturers depend on a global network of suppliers for critical components, such as semiconductor chips, powertrain parts, and specialized materials. Delays in sourcing these components can halt production, leading to inefficiencies and increased operational costs. Additionally, managing inventory for an assembly line requires precise coordination of thousands of parts arriving at the right time and location. Any misalignment in parts availability can lead to bottlenecks, impacting overall production flow. By integrating historical data, market trends, and real-time sensor-based monitoring (Mykoniatis, 2020), DTs can improve demand forecasting, ensuring better synchronization between parts supply and production schedules (Zaidi et al., 2024). This can help reduce both stockouts and excess inventory, improving overall efficiency.

While proprietary software platforms can provide visualization, collision control, and geometry accuracy while integrating robots from different manufacturers (Shtabel et al., 2024), open-source approaches to DT implementation offer benefits like interoperability and cost reduction (Damjanovic-Behrendt & Behrendt, 2019). Various open-source tools and technologies are available for DT development, covering data management, modeling, and services (Kazala et al., 2021; Bonney et al., 2022). However, integration challenges remain for open-source solutions (Kazala et al., 2021). Commercial platforms like Microsoft Azure and Siemens Xcelerator are preferred for net-zero manufacturing, though open-source alternatives like Bentley iTwin.js are gaining traction (Datta et al., 2024). DT implementation faces challenges in creating accurate models, real-time communication, and developing analysis methods (Rolle et al., 2020). Research gaps include addressing technical limitations and realizing integrated, flexible manufacturing environments (Murat et al., 2022; Onaji et al., 2022).

Moreover, DTs allow manufacturers to simulate potential disruptions, such as supplier delays or machinery failures, and develop contingency plans to mitigate their impact. Digital Twins can identify early signs of equipment wear through predictive maintenance, allowing for proactive servicing and reducing unplanned downtime. This predictive capability ensures smooth assembly line operations, reduces production delays, and ultimately enhances productivity and cost efficiency in automotive manufacturing (van Dinter et al., 2022).

11.2.3 Agent-Based Digital Twin Models

Agent-based modeling (ABM) is increasingly used in manufacturing and related industries to simulate complex systems and support decision-making. In manufacturing, ABM has been applied to optimize scheduling, improve resource allocation, and reduce project delays (Saeidlou et al., 2019; Mazzetto, 2024). ABM is also valuable for exploring sustainability in manufacturing supply chains, often combined with other simulation techniques like Discrete Event Simulation (DES) and System Dynamics (SD) (Khan & Abonyi, 2022). The architecture, engineering, and construction industries have benefited from ABM in project management, safety protocols, and sustainability initiatives (Mazzetto, 2024).

The use of simulation modeling techniques to create DTs is becoming increasingly prominent in research, with a primary focus on DES and SD. However, ABM and hybrid multi-paradigm simulation models (Mykoniatis & Angelopoulou, 2020) remain underrepresented in this domain (dos Santos et al., 2022). ABM involves simulating the interactions of autonomous agents within a system (Mykoniatis, 2015). This approach is particularly well suited for modeling dynamic processes that depend on agent interactions, with the aim to optimize collective behavior and achieve specific system objectives (Macal & North, 2009). In the case where the modeling process focuses on the interaction and data exchange between entities, the nature of agent-based DTs can make them excel for adaptive decision-making in complex environments (Marah & Challenger, 2024).

11.2.4 Gaps in Existing Research

While the concept of DTs has been widely discussed in the context of supply chain optimization, there are still notable gaps in the existing literature, particularly in the specific area of automotive assembly lines. Much of the research on DTs in supply chains has focused on the manufacturing processes and logistics, with less emphasis on their application in the dynamic and highly automated environment of automotive production lines. Given the complexity of automotive assembly, which involves a high volume of parts, intricate sequencing, global sourcing, and the just-in-time (JIT) manufacturing model, the potential of DT technology to optimize these processes remains underexplored.

Additionally, while numerous studies highlight the theoretical benefits of DT technology, there is limited research on real-world case studies demonstrating practical implementation and measurable outcomes in automotive assembly lines. Integrating DTs with production planning, real-time monitoring, and predictive maintenance in automotive plants presents a significant opportunity for reducing downtime, improving efficiency, and enhancing quality control. Furthermore, there is a lack of research on tracking the vehicles as they move through the assembly process, as most works focus solely on monitoring the manufacturing processes and do not include the products.

Monitoring key performance indicators (KPIs) such as cycle times, efficiency of the workstation, defect rates, and resource utilization in real time can provide valuable insight to optimize production flow. However, current studies do not address how DTs can dynamically collect, analyze, and visualize these metrics to support data-driven decision-making. Integrating real-time tracking with DT-based analytics could enable proactive adjustments, improve synchronization between workstations, and improve overall production efficiency, an underexplored area in the existing literature.

This chapter aims to bridge some of these gaps by presenting a multi-agent-based DT creation framework and a case study on the real-world application of DTs in a simulated automotive assembly line. The case study highlights how DT technology can be used to support applications that improve various aspects of production. These applications include improving overall equipment effectiveness, inventory management, production scheduling, and eliminating waste.

11.3 MABDT Architecture

Against this background, we propose a high-level architecture for designing DTs for manufacturing supply chains. This architecture utilizes agent-based modeling techniques to create a multi-agent system that captures the interactions between important aspects of the assembly process.

The proposed architecture combines digitalization efforts into three layers. The first is the Physical Layer, which contains all processes that can provide data to the models and exist in the physical space. Examples of systems that can be included in the Physical Layer are PLCs, Automation and Robotics systems, sensors, IoT devices, and industrial machinery such as Computer Numerical Control (CNC) machines. The next is the Digital Twin Layer, where, in this layer, we include all the required programs that are responsible for creating the actual system's digital counterpart. This layer can include all communication components, the DT of the factory, simulation models, and data storage. Finally, the Application Layer includes all applications that can be created utilizing the data gathered and created by the

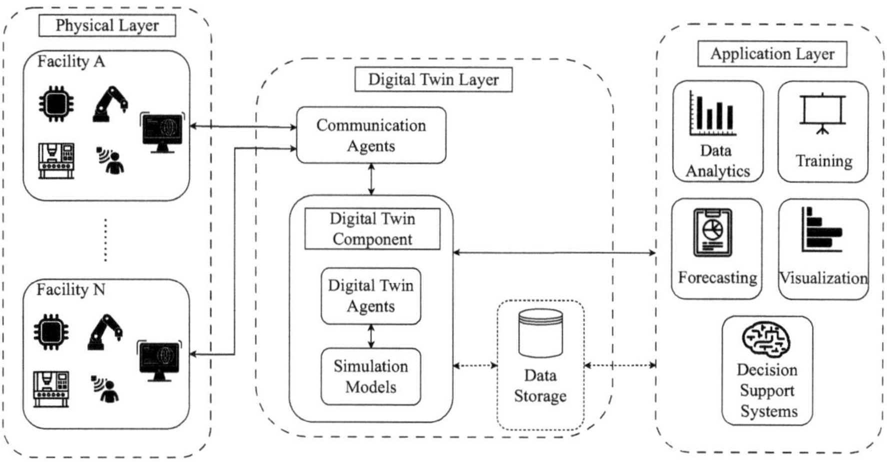

Fig. 11.1 A high-level overview of the components of the multi-agent-based Digital Twin model architecture

Digital Twin Layer. Examples of such applications could include data analytics, forecasting, training, visualization, and potential decision support systems.

As data is generated from the different facilities' assembly processes in the Physical Layer, it is made available to a collection of digital tools responsible for processing the incoming stream of information, keeping track of the system's state and eventually making this data available to valuable applications. We consider these digital tools as part of the Digital Twin Layer. Although some tools are not conventionally part of the DT, all components required for the DT model to operate successfully are classified as components of the Digital Twin Layer.

A high-level overview of this architecture with example inputs and outputs is illustrated in Fig. 11.1.

The first step in consuming the data incoming from the Physical Layer is to understand which specific manufacturing process generated the data and which agent model is responsible for the incoming data point. To achieve this, we employ Communication Agents, who work as data orchestrators tasked with transforming the incoming data point into a suitable message and propagating it to the correct part of the DT model.

The specific requirements to achieve communication between the Physical Layer and the Communication Agents are not formalized in this architecture, as in the real world, it heavily depends on equipment compatibility and operational requirements. For example, Modbus is a protocol often used in legacy systems to connect PLCs to sensors or actuators, suitable for smaller scale systems with straightforward data exchange needs. Another option is MTConnect, particularly for standardizing data collection from diverse manufacturing equipment, but it is limited to read-only operations, which hinders scenarios requiring control capabilities. Another way to establish bidirectional communication would be through a relational database, as some PLC models support direct reading and writing with popular open-source databases such as MySQL and PostgreSQL. The Open Platform Communications Unified Architecture (OPC UA), which is an industrial communication protocol that enables secure, platform-independent data exchange between machines, devices, and enterprise systems, is also well suited for environments requiring secure, scalable communication but may involve higher implementation complexity than other alternatives such as the Message-Queuing Telemetry Transport (MQTT) publish-subscribe messaging protocol.

The Digital Twin Layer is responsible for processing, analyzing, and leveraging data from the Physical Layer. Within this layer, DT Agents receive and process messages routed by the Communication Agents and keep track of the actual system's state using statecharts. By doing so, DT Agents can update their internal logic and trigger necessary actions based on real-time conditions.

Once updated, DT Agents can send commands back to the Physical Layer (via the Communication Agents) to enact changes in machinery or processes. For example, if a DT Agent detects a fault in a production line, it can send a message back to the physical system to halt operations or reroute defective products for rework. This bidirectional communication ensures that the digital model mirrors and influences physical operations.

In addition to real-time interactions, DT Agents are integrated with Simulation Models within the Digital Twin Layer. These models allow for predictive analysis by simulating future operational scenarios based on current system states. For instance, simulation models can assess how changes in production schedules or equipment configurations may impact throughput or lead times. Subsequently, DT Agents can recommend or directly implement proactive adjustments to optimize system performance. Simulation outputs, such as system performance predictions under various conditions, can directly inform applications like scenario planning and optimization tools.

The processed information from DT Agents and Simulation Models is either stored in a Data Storage System or made accessible to the Application Layer directly. The data storage can serve as a historical and real-time data repository and enables long-term analysis and traceability. The data processed within the Digital Twin Layer can be made available to the Application Layer through multiple mechanisms. Similar to the connection between the Physical and Digital Twin Layers, this architecture does not formalize the connection to the Application Layer. The Application Layer's way of receiving and transmitting data must be tailored to specific operational needs.

Real-time data streaming is one practical approach, where protocols like MQTT or WebSocket enable real-time, bidirectional data exchange updates from the DT Agents or simulation models to the applications. This is particularly beneficial for time-sensitive tasks such as predictive maintenance or fault detection. Another standard method involves storing processed data in centralized or distributed databases. Applications in the Application Layer can query these repositories for historical or aggregated data, enabling retrospective analysis and long-term performance evaluations. On the side of API-based communication, RESTful APIs, which follow a stateless, client-server architecture using standard HTTP methods (GET, POST, PUT, DELETE), provide structured access to processed data, while GraphQL APIs offer a more flexible alternative by allowing clients to request only the specific data they need, minimizing over-fetching and under-fetching of information. These APIs allow external applications to request specific datasets or trigger simulations on demand. Event-driven architectures, where applications react to events or changes in system states rather than relying on direct requests, often employ message brokers like Apache Kafka or RabbitMQ. Apache Kafka, a distributed streaming platform, enables high-throughput, fault-tolerant messaging and real-time data streaming, while RabbitMQ, a message-queuing broker, focuses on reliable message delivery and can be used for managing complex routing patterns in messaging systems.

The Application Layer leverages processed data from the Digital Twin Layer to support various applications. Data analytics is one key application area, enabling insights into trends, patterns, and anomalies through historical and real-time data analysis. Forecasting involves predictive models, including traditional statistical methods and Machine Learning (ML) models, built on historical and real-time data to help organizations anticipate future production outcomes, such as potential bottlenecks or equipment failures.

Training is also a significant use case; animating manufacturing processes in a simulated environment generated by the DT can be used to train personnel in a risk-free environment without disrupting actual operations. Visualization tools offer graphical representations of system states and performance metrics, helping operators and managers understand outputs and improve situational awareness. Finally, Decision Support Systems (DSSs) integrate analytical tools with real-time feedback from the Digital Twin Layer to assist managers in making informed decisions. Together, these applications illustrate how the Application Layer acts as a bridge between raw data processing in the Digital Twin Layer and useful insights that drive improved performance across manufacturing supply chains. With the consideration of the Application Layer as a separate entity in the architecture, we want to illustrate that DT technologies are not the end goal of a digitalization effort by itself. However, it can be an enabler technology to sandbox impactful applications that support the entire supply chain.

11.4 Digital Twin Agents

This section focuses on how the Digital Twin Agents' core behavioral logic and interactions with other agents are created using basic structural elements. A critical aspect of implementing MABDTs supply chain management is establishing a straightforward approach to defining agent behavior and communication interactions. The effectiveness of a DT system largely depends on how accurately its constituent agents represent their physical counterparts and how efficiently they exchange information within the digital environment. Statecharts provide a formal, visual mechanism for modeling the complex states and transitions that characterize agent behavior throughout various operating conditions. By standardizing the way these statecharts are constructed and interpreted, developers can ensure consistency across different agent types while facilitating system maintenance and scalability as the DT evolves to accommodate new supply chain elements.

Within a multi-agent DT environment, clearly defined communication is essential to maintain synchronization between the agents and enable coordinated responses. This communication framework must support various types of interactions, including intra-agent messages (within the same agent), intra-model exchanges (between different agents in the Digital Twin Layer), and external communications with the Physical and Application Layers. The present work focuses specifically on how DT agents are created, how they monitor processes and products within the automotive manufacturing supply chain, and how they communicate with each other to maintain an accurate representation of the physical system. The exploration of simulation models and their points of interaction with DT agents falls outside the scope of this current work, as research is ongoing to determine optimal integration methods between real-time multi-agent-based representations and predictive simulation models.

11.4.1 Defining the Digital Twin Model's Agents

The structural elements required to create and program the agents are illustrated in Fig. 11.2, where we showcase the structural elements to define agent statecharts and their communication mechanisms.

The design of agents in the MABDT model relies on statecharts to define their behavior, transitions, and interactions. Statecharts provide a structured framework for modeling each agent's life cycle, capturing its states, transitions, and responses to internal or external triggers. The construction of these statecharts begins with an Initial State Indicator, marking the starting point of an agent's behavior when it is instantiated. Each state represents a specific phase or condition in the agent's life cycle, while Composite States allow for hierarchical modeling by grouping multiple substates under a higher level state.

Transitions between states are governed by two primary mechanisms, Conditional Transitions and Synchronization Transitions. Conditional transitions occur when predefined conditions are met, such as when a counter reaches a threshold or production time elapses. Synchronization transitions, on the other hand, are triggered by messages received from Communication Agents. These transitions specify the target state and the message that initiates the transition. The life cycle concludes with a Final State Indicator, marking the end point of an agent's process.

Two unique indicators are used, when needed, to manage the transition into a composite state: Retain Composite State History (RH) and Do Not Retain Composite State History (NH). The RH indicator ensures that when an agent reenters a composite state, it resumes from the substate it previously existed in when

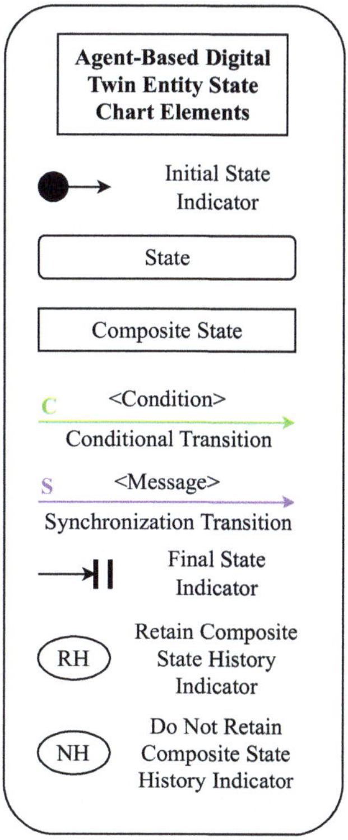

State Communication Interfaces
State1: • IncDataPoint1 : Intra-agent - Incoming • OutDataPoint2 : Intra-model - Outcoming • IncDataPoint3 : External - Incoming
State2: • IncDataPoint1 : Intra-agent - Incoming • OutDataPoint2 : Intra-model - Outcoming • IncDataPoint3 : External - Incoming

Agent Interaction List
Incoming Communication: • Agent1 : IncDataPoint1 • Agent2 : IncDataPoint2
Outcoming Communication: • Agent1 : OutDataPoint1 • Agent2 : OutDataPoint2

State Action List
State1: • Function1(Inputs): Synch - Return Value - On Enter • Function2(Inputs): Asynch - Return Value - On Exit
State2: • Function1(Inputs): Synch - Return Value - On Enter • Function2(Inputs): Asynch - Return Value - On Exit

Agent Containing Elements
+ prop1: property type
+ var1: variable type
+ agent1: containing agent type
+ RecurringEvent1: Time Interval (unit)

Fig. 11.2 The basic elements of the agent-based statecharts, the State Communication Interfaces Table, the State Action List Table, the Agent Interaction List Table, and the Agent Containing Elements Table

it was in this composite state previously. In contrast, the NH indicator resets the agent to the substate marked as its initial entry point within that composite state.

In addition to defining state behaviors, agents also manage their communication interfaces. Each state is associated with specific data points that define incoming and outgoing interactions. These data points are categorized based on their origin or destination. Intra-agent communication occurs within the same agent, intra-model communication involves other noncommunication agents, and external communication refers to messages sent from and to the Physical Layer via Communication Agents.

The interaction between agents is further detailed in an Agent Interaction List, specifying all agent pairs involved in communication. This includes incoming messages from other agents and outgoing messages sent to them.

Each state may also require specific actions upon entry or exit, which are defined in a State Action List. These actions are implemented as functions or code snippets that execute during these transitions. Functions can be synchronous, requiring the agent to wait for completion before proceeding, or asynchronous, allowing other tasks to continue while awaiting results.

Finally, agents are characterized by their internal properties and hierarchical structure through an Agent Containing Elements list. This includes static property values that define key attributes of an agent and dynamic variables used during runtime for computations or tracking internal states. Additionally, some agents may contain subagents, such as a subcell agent managing workstations within its scope. Recurring events are also defined here as periodic actions triggered at regular intervals to execute specific functions.

By integrating these structural elements (statecharts, communication interfaces, interaction lists, action lists, and containing elements), agents in the ABDT system achieve a high degree of modularity and adaptability. This approach facilitates robust modeling of complex behaviors and focuses on scalability and interoperability within dynamic manufacturing environments.

11.4.2 Communication Between Agents

In the MABDT architecture, selecting the number of communication agents and the location in the network of the DT Agents can be tailored to adopt one of four multi-agent system (MAS) architectures (Centralized, Hierarchical, Heterarchical, or Distributed), as outlined in the work of Salvador Palau et al. (2019).

In a Centralized architecture, all decision-making occurs at a central node. In this setup, Digital Twin Agents are all hosted in a central location and primarily collect and transmit data to a central Communication Agent. While this simplifies communication and reduces communication complexity, it introduces a single point of failure and limits scalability in large systems.

A Hierarchical architecture considers each facility to have its own Communication Agent. These agents manage subsets of Digital Twin Agents, handling localized updates and communicating with the Communication Agents of other facilities when needed. This approach balances communication efficiency and scalability but may introduce latency due to the sequential flow of information.

In a Heterarchical architecture, Digital Twin Agents can communicate directly with each other while still being loosely coordinated by a central Communication Agent that handles external communication from the Physical Layer. This setup allows agents to exchange information about shared resources or dependencies, such as material shortages or production delays. Peer-to-peer interactions improve adaptability and responsiveness but require robust mechanisms to avoid conflicts or redundancies.

Finally, a Distributed architecture would eliminate centralized control by Communication Agents entirely, enabling all DT Agents to operate autonomously with peer-to-peer communication. Each agent independently manages its statechart while coordinating directly with others. This structure is highly resilient to failures and well suited for complex supply chains. However, it demands sophisticated algorithms for distributed decision-making to ensure system-wide consistency.

The choice of architecture depends on operational priorities such as scalability, fault tolerance, and communication efficiency. Centralized or hierarchical models may be better suited for smaller scale systems or those requiring centralized oversight. In contrast, heterarchical or distributed architectures are ideal for dynamic environments requiring high adaptability and robustness against disruptions.

11.5 Case Study

To illustrate the potential of Digital Twins in manufacturing, we present a case study of the Tiger Motors Assembly Line. Tiger Motors is an educational facility located at the Samuel Ginn College of Engineering at Auburn University. It simulates a high-volume automotive manufacturing plant and serves as a platform for teaching Lean Manufacturing (LM) principles and providing hands-on training to students, industry professionals, and organizations. The facility emulates an automotive supply chain environment, offering a comprehensive view of production processes, material handling, and inventory management. Tiger Motors operates under a target takt time of 75 seconds (calculated by dividing available production time by customer demand) and a target cycle time of 60 seconds, close to real-life automotive manufacturers' demand and production rates.

11.5.1 The Tiger Motors Assembly Line

The Tiger Motors assembly line comprises three subassembly cells, each containing five workstations. Production occurs in a mixed-mode operation, where two different car types—a 234-piece SUV and a 277-piece Speedster—are assembled concurrently on the same line made from LEGO®bricks. The assembly line is divided into three subassembly cells, Cell 1, Cell 2, and Cell 3, each comprising five workstations (WS1–WS15). Cars move sequentially through these workstations during their assembly process.

At the end of each subassembly cells, there is an inspection station to ensure product quality before cars proceed further down the line. Additionally, inventory supermarkets are strategically placed at the beginning of Subassembly Cells 2 and 3 to store incomplete cars temporarily. These supermarkets facilitate smooth transitions between subassembly cells while minimizing delays caused by production imbalances. The facility also incorporates material handling systems using a two-bin replenishment system at each workstation. Material handlers restock these bins from a material supermarket area, replenished by an automatic storage and retrieval system (ASRS). The ASRS is conceptualized as a truck delivering materials from external suppliers.

Beyond the main assembly line, Tiger Motors utilizes an independent manufacturing cell known as War Eagle Inc., which produces specialized add-ons for modified car orders. This cell operates as a five-step manufacturing process synchronized with the cycle time of the main assembly line. When a modified car enters workstation 11 (WS11), an order is triggered to War Eagle Inc., ensuring that the customized part is ready by the time the car reaches WS15, the final workstation.

Production at Tiger Motors is structured to simulate a real-world supply chain environment. A process flow of Tiger Motors' supply chain is illustrated in Fig. 11.3. In Fig. 11.4 we illustrate key elements of the Tiger Motors educational facility: (i) Cell 3, the final subassembly cell, featuring workstations 11–15 with a continuously moving conveyor belt where specialized add-ons from War Eagle Inc. are integrated into modified vehicles (top left), (ii) the inspection station at the end of the assembly line (bottom left), (iii) the standardized work and scanning instructions displayed at a workstation, guiding operators through proper assembly procedures and barcode scanning protocols used for real-time production tracking (top right), and (iv) a typical workstation work area showing the two-bin material management system, where operators assemble components and scan completed work, triggering LED indicators that provide visual confirmation of proper scanning operations (bottom right).

The production begins in Cell 1 at WS1, where operators assemble cars using materials organized in two-bin systems at each workstation. When one bin empties, a replenishment request is sent to material handlers, who restock it from the material supermarket area. This area is replenished by an Automatic Storage and Retrieval System (ASRS) implementation that mimics external supplier deliveries. As cars progress through Cell 1, they undergo incremental assembly steps until they reach WS5, where they are inspected for quality issues. Any defective cars are either reworked or removed from the system based on their condition.

After passing inspection in Cell 1, cars enter Supermarket 1 (SM1), an intermediate inventory storage area. SM1 batches incomplete cars before transferring them to Cell 2 for further assembly. In Cell 2 (WS6–WS10), cars continue along similar processes with additional components added at each workstation. At WS10, another inspection station ensures quality before cars are transferred to Supermarket 2 (SM2). This second inventory storage area serves as a buffer before cars enter Cell 3.

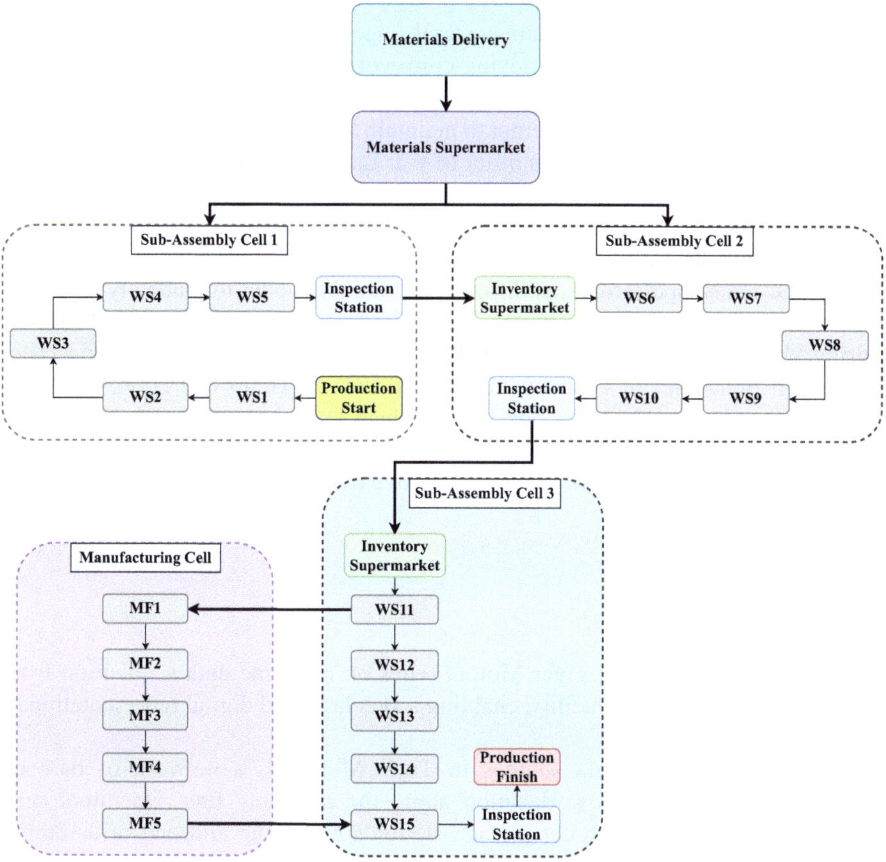

Fig. 11.3 The process flow of the Tiger Motors supply chain

Fig. 11.4 Examples of the Tiger Motors assembly line stages of production

Cell 3 (WS11–WS15) represents the final stage of assembly and includes unique features such as a continuously moving conveyor belt that transports cars between workstations unless halted by an emergency or defect detection. Inspections occur at WS5, WS10, and WS15 as an attempt to maintain high product quality standards. At WS11, car arrivals may trigger an order to War Eagle Inc., initiating the production of specialized add-ons required for customization. By the time these modified cars reach WS15, their corresponding add-ons are delivered from War Eagle Inc., completing their assembly.

The line can be operated under different production modes to gradually introduce operators to the transition from the Mass Production paradigm to an LM one. During Mass Production runs, the facility operates according to the outlined Master Production Schedule. In contrast, under LM (Just-In-Time) operation, a leveled schedule based on a Heijunka box helps reduce lead time and work-in-progress (WIP) inventory. By simulating both modes of operation, operators learn how Lean principles can reduce inefficiencies and improve overall system performance.

11.5.2 Data Sources

The DT implementation at Tiger Motors relies on real-time data acquisition from various sources within the facility, enabling a synchronized digital representation of the physical assembly line.

One of the primary data sources in Tiger Motors is a network of barcode scanners installed at each workstation along the assembly line. Operators scan barcodes placed under each car at each station's beginning and end of assembly tasks. This process enables real-time tracking of car locations and provides insights into whether workstations are busy or idle. Each barcode scanner is connected to a Raspberry Pi microcomputer, which acts as an IoT device. These devices wirelessly transmit scanned data to an Eclipse Mosquitto MQTT broker hosted on a central computer, located in the robot cage at the end of the production line. The MQTT broker enables communication by publishing data streams that can be subscribed to by various applications, including those used in the DT model.

The barcode scanning system also helps with operator training and ensures they adhere to assembly and data collection protocols. For instance, when a car is scanned for the first time at a workstation, the corresponding workstation agent in the digital twin sends an MQTT message back to the IoT device to activate an LED indicator. This visual cue reminds operators to complete their scanning tasks correctly. Upon scanning the car again at the end of the assembly, another message is sent to turn off the LED, signaling task completion.

In addition to tracking cars, barcode scanners are used for material handling operations. Replenishment bins at each workstation are equipped with distinct barcodes. When a bin is emptied and scanned by an operator, a replenishment request for material handlers is triggered. Then, the material handlers restock it from the material supermarket area. If a part needs to be retrieved from storage, scanning a code in the material supermarket triggers a process for the ASRS to collect and deliver materials to the supermarket.

Another key Physical Layer data source is Tiger Motors' automation system. Controlled by an Omron NJ101 PLC, the automation system manages various components, including inspection cameras, Andon lights and control buttons, a robotic sorting application at the end of the production line, and conveyor belt operations in Cell 3. To facilitate communication, KEPserverEX, an industrial connectivity platform, exposes the PLC tags and variables via an OPC UA server to the network. This setup leverages OPC UA's publish-subscribe capabilities, ensuring data exchange happens only when changes occur in the automation system. While SQL-based communication is an alternative, it requires continuous polling by communication agents, making it less computationally efficient than OPC UA's event-driven approach.

Integrating these data sources into Tiger Motors' DT framework supports additional supply chain functionalities. For example, when a car requiring extra modifications is scanned at WS11 in Cell 3, a message is sent to the War Eagle Inc. external independent manufacturing line. This ensures that when cars that require add-ons from War Eagle Inc. reach WS15 (the final workstation) the add-on parts are delivered and ready to be installed.

11.5.3 Defining Agents for the Tiger Motors Digital Twin Model

The development of a DT for Tiger Motors involves defining and programming agents that represent various elements of the facility's supply chain. In the Tiger Motors case study, the agents are implemented using AnyLogic© 8.8.6 simulation modeling software, with real-time communication with the Physical Layer achieved via MQTT. AnyLogic enables the creation of agent-based statecharts and was previously used to develop a simulation model of the Tiger Motors facility (Katsigiannis et al., 2024). This prior work considered the creation of a hybrid simulation model of the facility, examining the effects of transitioning from the Mass Production paradigm to the LM paradigm.

The DT architecture for this model employs a centralized communication approach, where one Communication Agent acts as the orchestrator for data exchange between IoT devices, automation systems, and other agents within the DT layer. A more detailed review of how the agents were developed and the data collection protocols that were followed can be found in our prior work (Katsigiannis & Mykoniatis, 2024).

The Communication Agent serves as the central hub for receiving, processing, and distributing data from the physical system to the respective DT Agents. It subscribes to MQTT topics published by IoT devices, such as the barcode scanners at workstations, and parses incoming messages to identify their origin and purpose.

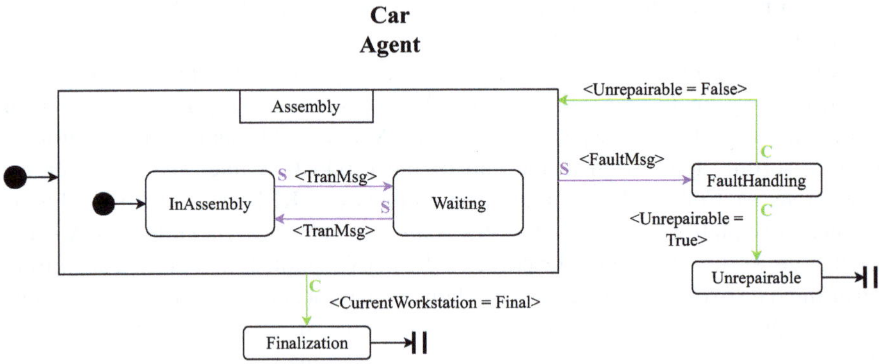

Fig. 11.5 The Car Agent's statechart

For example, when a car is scanned at a workstation, the Communication Agent determines whether it corresponds to a new car entering the system or an existing car transitioning between stations. Based on this information, it either creates a new Car Agent or updates the state of existing Car and Workstation Agents. Additionally, it interacts with the Omron NJ101 PLC via KEPserverEX, using OPC UA to monitor and update automation-related data such as conveyor belt status or Andon light signals. This centralized architecture simplifies communication management but requires robust processing capabilities to handle real-time data flows effectively.

The Car Agent models each vehicle being assembled in Tiger Motors. Its behavior is governed by a statechart (illustrated in Fig. 11.5), which tracks the car's progression through the assembly line. The agent starts in an initial state when it is created upon scanning at WS1. It transitions through states representing its location at various workstations (e.g., "InAssembly" or "Waiting") based on messages from the Communication Agent. If a fault is detected during assembly, the agent enters a "FaultHandling" state, where it records defect details and determines whether rework or removal from the system is necessary. Upon completing all 15 workstations, the agent transitions to a "Finished" state, where key performance indicators such as lead time and defect statistics are recorded.

The Workstation Agent represents each of Tiger Motors' 15 workstations and monitors their operational states. Its statechart (shown in Fig. 11.6) includes states corresponding to Andon light statuses ("Normal Operation," "SupervisionAssistanceRequired," and "ProductionHalt"). Within the "NormalOperation," additional substates track whether the workstation is idle or busy based on interactions with Car Agents.

The Workstation Agent also interacts with other elements of the DT system. For example, when a modified car requiring an add-on part is scanned at WS11, it triggers an order to War Eagle Inc., initiating production of the customized component. This ensures that its add-on is ready for installation by the time the car reaches WS15.

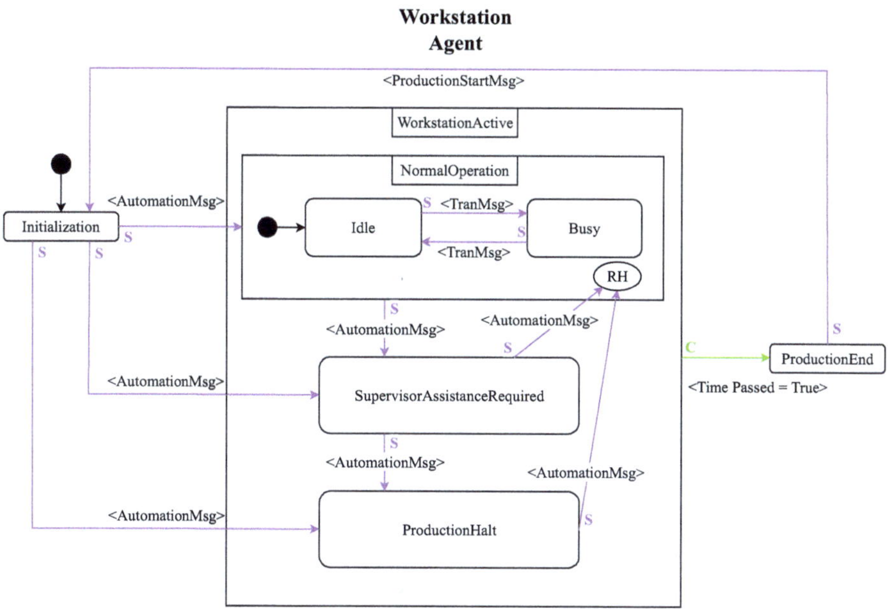

Fig. 11.6 The Workstation Agent's statechart

Workstation Agents also execute specific actions upon entering or exiting states. For example, when transitioning to a “Busy" state, they update utilization metrics and communicate with material handlers if replenishment bins are empty. Messages from barcode scanners or automation systems trigger these transitions via the Communication Agent. At the end of each production run, Workstation Agents transition to a “ProductionFinished" state to aggregate performance metrics such as service times and downtime.

War Eagle Inc., an external independent manufacturing line producing customized add-ons for modified cars, is represented by its agent in the DT system. This agent operates based on a five-step manufacturing process synchronized with Tiger Motors’ cycle time. When a modified car enters WS11 and triggers an order, this agent signals the beginning of production for the required component. It tracks progress through its internal statechart and monitors the delivery of add-ons to WS15 for final assembly.

Material handling operations are modeled through agents representing supermarkets and delivery systems. These agents track inventory levels using data from barcode scans performed during replenishment activities. When a bin barcode is scanned in the material supermarket area, it triggers an order to the ASRS system. The ASRS system then retrieves materials from storage and restocks the supermarket.

By integrating these agents into a centralized communication framework supported by real-time data acquisition systems, Tiger Motors’ Digital Twin achieves

State Action List
Assembly:
InAssembly: • CheckIfFinishingWorkstation(): Synch - Boolean - On Exit • InformWorkstation(): Asynch - None - On Enter
Waiting: • InformWorkstation(): Asynch - None - On Enter
FaultHandling: • UpdateListOfFaults(FaultType): Synch - None - On Enter • UpdateRepairable(FaultType): Synch - None - On Enter
Finalization: • UpdateStatistics(): Asynch - None - On Enter • RemoveCarAgent(): Asynch - None- On Exit
Unrepairable: • UpdateStatistics(): Asynch - None - On Enter • RemoveCarAgent(): Asynch - None- On Exit

Agent Interaction List
Incoming Communication: • Communication Agent: TranMsg • Communication Agent: FaultMsg
Outcoming Communication: • Workstation Agent: TranMsg

Agent Containing Elements
+ CarVIN: String + AgentCreationTimestamp: Timestamp
+ CurrentWorkstation: Workstation Agent + Unrepairable: Boolean + FaultsIdentified: List of Values - <Workstation, FaultType>
+
+

State Communication Interfaces
Assembly: • FaultMsg: External - Incoming
InAssembly • TranMsg: Inter-model - Incoming
Waiting: • TranMsg: Inter-model - Incoming
FaultHandling:
Finalization:
Unrepairable:

Fig. 11.7 The Car Agent's State Communication Interfaces Table, State Action List Table, Agent Interaction List Table, and Agent Containing Elements Table

high fidelity in modeling its supply chain operations. Figures 11.7 and 11.8 illustrate the State Action, Agent Containing Elements, Agent Interaction, and State Communication Interfaces lists for the Car and Workstation agents.

11.6 Conclusion

This chapter presented a DT architecture focused on the digitalization of automotive supply chains. The architecture employs a three-layer structure, comprising a Physical Layer, a Digital Twin Layer, and an Application Layer, which can integrate real-time data from IoT devices, automation systems, and Programmable Logic Controllers (PLCs). A comprehensive case study is also presented on the implementation of the MABDT architecture to digitalize an educational facility that simulates an automotive manufacturing supply chain. Utilizing MQTT protocols

State Action List
WorkstationActive:
NormalOperation:
Idle: • UpdateLEDOn(WSId): Synch - None - On Enter
Busy: • UpdateLEDOff(WSId): Synch - None - On Enter • OrderAddon(WSId): Asynch - None - On Enter
SupervisorAssistanceRequired:
ProductionHalt:
Initialization: • CollectStartingState(WSId): Synch - None - On Enter
ProductionEnd: • FinalizeStatistics(): Synch - None - On Enter

Agent Interaction List
Incoming Communication: • Car Agent: TranMsg • Communication Agent: AutomationMessage • Communication Agent: ProductionStartMessage
Outcoming Communication: • Communication Agent: UpdateLEDMsg • WarEagleInc Agent: OrderMsg

Agent Containing Elements
+ WSId: String
+ CurrentWorkstation: Workstation Agent + TotalBusyTime: Integer + TotalIdleTime: Integer + TotalDownTime: Integer + ProductionStartTime: Timestamp
+
+ UpdateBusyIdleAndDownTimes : 100 milliseconds

State Communication Interfaces
WorkstationActive: :
NormalOperation: • AutomationMessage: External - Incoming
Idle: • TranMsg: Inter-Model - Incoming • UpdateLEDMsg: External - Outcoming
Busy: • TranMsg: Inter-Model - Incoming • UpdateLEDMsg: External - Outcoming • OrderMsg: Inter-Model - Outcoming
SupervisorAssistanceRequired: • AutomationMessage: External - Incoming
ProductionHalt: • AutomationMessage: External - Incoming
Initialization: • AutomationMessage: External - Incoming • OrderMsg: External - Outcoming
ProductionEnd: • ProductionStartMessage: External - Incoming

Fig. 11.8 The Workstation Agent's State Communication Interfaces Table, State Action List Table, Agent Interaction List Table, and Agent Containing Elements Table

and OPC UA servers, this setup facilitated bidirectional communication between the physical and digital systems, enabling real-time tracking, synchronization with physical operations, and enhanced decision-making capabilities.

The implementation in the case study was developed using multi-paradigm simulation software, with agents representing key elements of the supply chain. These elements include cars, workstations, material supermarkets, and an external manufacturing cell. The agents are modeled using statecharts and structured communication interfaces, demonstrating the potential of Agent-Based Digital Twins in adaptive supply chain management. The case study highlighted the capabilities of the MABDT framework in modeling supply chain dynamics to monitor the automotive production process, while enabling essential applications such as takt time management, predictive maintenance, and automated material handling.

Beyond its application to automotive manufacturing, this framework can serve as a scalable blueprint for the integration of DTs across various industries.

11.6.1 Future Research Directions

The MABDT architecture presented in this chapter opens several promising research directions. Integrating DT agents with industrial simulation models would enhance decision support capabilities, enabling real-time monitoring and predictive scenario evaluation for proactive supply chain management. The Tiger Motors case study demonstrates implementation in a controlled environment, but scaling to complex manufacturing facilities with extensive supplier networks requires further research into optimizing agent communication patterns and data processing efficiency.

Examining alternative agent architectures—distributed and heterarchical versus our centralized approach—could yield valuable insights into performance trade-offs for different operational contexts. As these systems evolve toward greater autonomy, robust verification and validation methodologies will become increasingly important, particularly in safety-critical environments.

Further research should explore the adaptability of DT solutions across different industries with varying supply chain complexities. Additionally, addressing ethical, regulatory, and security considerations will be crucial as more sensitive data is integrated into these systems. While DT technologies have already demonstrated a significant impact in automotive manufacturing, continued research will help unlock their full potential for transforming supply chain management across industries.

References

BMW Group. (2022). BMW iFactory digital: Artificial intelligence in production. https://www.bmwgroup.com/en/news/general/2022/bmw-ifactory-digital.html. Accessed 20 Feb 2025.

Bonney, M.S., de Angelis, M., Dal Borgo, M., et al. (2022). Development of a digital twin operational platform using python flask. *Data-Centric Engineering, 3*. https://doi.org/10.1017/dce.2022.1

Damjanovic-Behrendt, V, & Behrendt, W. (2019). An open source approach to the design and implementation of digital twins for smart manufacturing. *International Journal of Computer Integrated Manufacturing*. https://doi.org/10.1080/0951192X.2019.1599436

Datta, P., Sharma, G., Anand, N., et al. (2024). A comparative analysis for harnessing digital twin platforms for net-zero manufacturing. *IEEE Access*. https://doi.org/10.1109/ACCESS.2024.3447475

Dong, L., & Kouvelis, P. (2020). Impact of tariffs on global supply chain network configuration: Models, predictions, and future research. *Manufacturing & Service Operations Management, 22*(1), 25–35.

dos Santos, C.H., Montevechi, J.A.B., de Queiroz, J.A., et al. (2022). Decision support in productive processes through des and abs in the digital twin era: a systematic literature review. *International Journal of Production Research, 60*(8), 2662–2681.

Financial Times. (2025). European carmakers brace for a deeper and longer downturn. https://www.ft.com/content/7e0ca35e-d6a8-48b7-b657-cf547f3994cb. Accessed: 20 Feb 2025.

Ford Motor Company. (2023). Ford Motor Company Dearborn Campus uses digital twin tool in energy plant. https://betterbuildingssolutioncenter.energy.gov/implementation-models/ford-motor-company-dearborn-campus-uses-digital-twin-tool-energy-plant. Accessed: 20 Feb 2025.

Grieves, M. (2024). Digital twins and their role in reengineering engineering education. In *Digital twin* (pp. 237–261). Springer.

Katsigiannis, M., & Mykoniatis, K. (2024). Real-time tracking of production in assembly operations using agent-based modeling and digital twin techniques. In *2024 Winter Simulation Conference (WSC)* (pp. 3011–3021). IEEE.

Katsigiannis, M., Pantelidakis, M., & Mykoniatis, K. (2024). Assessing the transition from mass production to lean manufacturing using a hybrid simulation model of a LEGO® automotive assembly line. *International Journal of Lean Six Sigma, 15*(2), 220–246.

Kazala, R., Luściński, S., Straczyński, P., et al. (2021). An enabling open-source technology for development and prototyping of production systems by applying digital twinning. *Processes*. https://doi.org/10.3390/pr10010021

Khan, A.A., & Abonyi, J. (2022). Simulation of sustainable manufacturing solutions: tools for enabling circular economy. *Sustainability*. https://doi.org/10.3390/su14159796. https://www.mdpi.com/2071-1050/14/15/9796/pdf?version=1660113403

Le, T.V., & Fan, R. (2024). Digital twins for logistics and supply chain systems: Literature review, conceptual framework, research potential, and practical challenges. *Computers & Industrial Engineering, 187*, 109768.

Lugaresi, G., Jemai, Z., & Sahin, E. (2023). Digital twins for supply chains: Current outlook and future challenges. In *Communications of the ECMS*.

Macal, C.M., & North, M.J. (2009). Agent-based modeling and simulation. In *2009 Winter Simulation Conference (WSC)* (pp. 86–98). https://doi.org/10.1109/WSC.2009.5429318

Marah, H., & Challenger, M. (2024). Adaptive hybrid reasoning for agent-based digital twins of distributed multi-robot systems. *Simulation, 100*(9), 931–957. https://doi.org/10.1177/00375497231226436

Maware, C., & Parsley, D.M. (2022). The challenges of lean transformation and implementation in the manufacturing sector. *Sustainability, 14*(10), 6287.

Maware, C., Okwu, M.O., & Adetunji, O. (2022). A systematic literature review of lean manufacturing implementation in manufacturing-based sectors of the developing and developed countries. *International Journal of Lean Six Sigma, 13*(3), 521–556.

Mazzetto, S. (2024). Interdisciplinary perspectives on agent-based modeling in the architecture, engineering, and construction industry: a comprehensive review. *Buildings*. https://doi.org/10.3390/buildings14113480

Murat, A., Murat, U., Oksuz, B., et al. (2022). Digital twins in manufacturing: systematic literature review for physical–digital layer categorization and future research directions. *International Journal of Computer Integrated Manufacturing*. https://doi.org/10.1080/0951192X.2021.2022762

Mykoniatis, K. (2015). *A generic framework for multi-method modeling and simulation of complex systems using discrete event, system dynamics and agent based approaches* (Publication No. 1391) [Doctoral dissertation, University of Central Florida]. STARS. https://stars.library.ucf.edu/etd/1391

Mykoniatis, K. (2020). A real-time condition monitoring and maintenance management system for low voltage industrial motors using Internet-of-Things. *Procedia Manufacturing, 42*, 450–456.

Mykoniatis, K., & Angelopoulou, A. (2020). A modeling framework for the application of multi-paradigm simulation methods. *Simulation, 96*(1), 55–73.

Mykoniatis, K., & Harris, G.A. (2021). A digital twin emulator of a modular production system using a data-driven hybrid modeling and simulation approach. *Journal of Intelligent Manufacturing, 32*(7), 1899–1911.

Narula, S., Puppala, H., Kumar, A., et al. (2023). Are industry 4.0 technologies enablers of lean? evidence from manufacturing industries. *International Journal of Lean Six Sigma, 14*(1), 115–138.

Onaji, I., Tiwari, D., Soulatiantork, P., et al. (2022). Digital twin in manufacturing: conceptual framework and case studies. *International Journal of Computer Integrated Manufacturing.* https://doi.org/10.1080/0951192X.2022.2027014

Pantelidakis, M., & Mykoniatis, K. (2024). Simulation aspects of a generic digital twin ecosystem for computer numerical control manufacturing processes. In *2024 Winter Simulation Conference (WSC)* (pp. 1599–1610). IEEE.

Pantelidakis, M., Mykoniatis, K., Liu, J., et al. (2022). A digital twin ecosystem for additive manufacturing using a real-time development platform. *The International Journal of Advanced Manufacturing Technology, 120*(9), 6547–6563.

Reuters. (2025). Companies race to ship goods to the us ahead of potential tariffs. https://www.reuters.com/business/companies-race-ship-goods-us-ahead-potential-tariffs-2025-01-30/. Accessed 20 Feb 2025.

Rolle, R.P., Martucci, V.d.O., & Godoy, E.P. (2020). Architecture for digital twin implementation focusing on industry 4.0. *IEEE Latin America Transactions.* https://doi.org/10.1109/TLA.2020.9082917

Saeidlou, S., Saadat, M., Amini Sharifi, E., et al. (2019). Agent-based distributed manufacturing scheduling: An ontological approach. *Cogent Engineering, 6*(1), 1565630.

Salvador Palau, A., Dhada, M.H., & Parlikad, A.K. (2019). Multi-agent system architectures for collaborative prognostics. *Journal of Intelligent Manufacturing, 30*(8), 2999–3013.

Shtabel, N., Razumovsky, A., Tkachev, S., et al. (2024). Digital twin of production on the assembly line. In *Proceedings of the III International Conference on Advances in Science, Engineering, and Digital Education: ASEDU-III 2022*. AIP Publishing.

Stroumpoulis, A., & Kopanaki, E. (2022). Theoretical perspectives on sustainable supply chain management and digital transformation: A literature review and a conceptual framework. *Sustainability, 14*(8), 4862.

Tiwari, M.K., Bidanda, B., Geunes, J., et al. (2024). Supply chain digitisation and management.

van Dinter, R., Tekinerdogan, B., & Catal, C. (2022). Predictive maintenance using digital twins: A systematic literature review. *Information and Software Technology, 151*, 107008.

Wang, L., Deng, T., Shen, Z.J.M., et al. (2022). Digital twin-driven smart supply chain. *Frontiers of Engineering Management, 9*(1), 56–70.

Zaidi, S.A.H., Khan, S.A., & Chaabane, A. (2024). Unlocking the potential of digital twins in supply chains: A systematic review. *Supply Chain Analytics*, 100075.

Zavala-Alcívar, A., Verdecho, M.J., Alfaro-Saiz, J.J. (2020). A conceptual framework to manage resilience and increase sustainability in the supply chain. *Sustainability, 12*(16), 6300.

Chapter 12
Digital Twins for Hyperconnected City Logistics

Russell G. Thompson, Masoud Kahalimoghadam, Lele Zhang, and Soheil Sabri

Abstract Rapid urbanization and the accelerated demand for e-commerce have resulted in a substantial increase in deliveries in many cities. There is an urgent need to reduce emissions from freight vehicles in urban areas to lessen climate change and improve the health of citizens.

Hyperconnected City Logistics (HCL) combines the concepts of City Logistics (CL) and the Physical Internet (PI). City logistics involves implementing solutions for minimizing the social and environmental impacts of freight in cities while maximizing the economic benefits of goods distribution in urban areas. The Physical Internet transforms independent freight systems into more open, collaborative, and integrated networks. This involves shared use of resources and facilities, including freight vehicles and warehouses.

PI networks involve the transshipment of goods at terminals (PI hubs) to increase network efficiency by improving consolidation levels of vehicles. The capacity and performance of vehicles can be matched to demand to promote a range of vehicle types to be utilized. This can increase the viability of alternative fuel vehicles and micromobility for last-mile distribution.

Digital Twins provide an effective modeling and simulation platform for facilitating the implementation of HCL, including assisting the exchange of goods at PI hubs to increase levels of coordination and enhancing the utilization of loading space, which can reduce vehicle delays and improve the system's reliability. Specifically, spatially enabled Digital Twins enable integration and harmonization of multi-dimensional data (2D, 3D, real-time). These capabilities, integrated with

R. G. Thompson (✉) · M. Kahalimoghadam
Faculty of Engineering and IT, The University of Melbourne, Melbourne, Australia
e-mail: rgthom@unimelb.edu.au

L. Zhang
School of Mathematics and Statistics, The University of Melbourne, Melbourne, Australia

S. Sabri
School of Modeling, Simulation and Training, University of Central Florida, Orlando, FL, USA

G. Rabadi, B. Soykan (eds.), *Optimizing Supply Chains Through Digital Twins*,
International Series in Operations Research & Management Science 375,
https://doi.org/10.1007/978-3-032-08147-6_12

modeling and simulation methods, will provide a robust decision support tool for the optimization of HCLs.

The synergy of PI, CL, and Digital Twins promises to revolutionize goods transportation through interconnected and efficient networks. This optimizes goods flow, providing real-time assessments of factors like traffic patterns, delivery routes, and hub utilization affecting transportation efficiency. Embracing this approach provides stakeholders with insights into the interconnected nature of city logistics, fostering a sustainable and responsive supply chain.

This chapter describes how spatially enabled Digital Twins can be developed to improve the performance of urban distribution systems incorporating micro-consolidation centers, loading zones, and intermodal terminals (rail and road).

Keywords City logistics · Physical Internet · Urban logistics · Spatial digital twins

12.1 Introduction

Cities generate a substantial amount of goods movement that is crucial for the well-being of residents and the economy. Urban freight systems facilitate access to markets, which is necessary for the production and consumption of goods. Emissions are rising due to the growth in urbanization and congestion levels, as well as the growth in e-commerce, which is increasing distances travelled by delivery vehicles.

Currently, a large proportion of truck and van trips in urban areas utilize only a small proportion of their load carrying capacity, and there is a substantial amount of empty travel. Consolidation has been identified as a key factor in decarbonizing freight (ITF, 2018). A range of analytical procedures are required to improve the efficiency of the urban freight networks. Descriptive analytics can be used to aid understanding of what happened in the past. This can include techniques for detecting trends, patterns, and incidents. Predictive analytics provide estimates of future events and conditions. Agent-based modeling can be used to simulate the interactions between various agents, such as shippers and carriers, in urban freight systems. Prescriptive analytics, such as optimization procedures, provide information on what the best values of control variables are to achieve desired outcomes.

Hyperconnected City Logistics (HCL) combines the concepts of City Logistics and the Physical Internet (Crainic & Montreuil, 2016). City logistics involves designing and implementing solutions for minimizing the social and environmental impacts of freight in cities while maximizing the economic benefits of goods distribution in urban areas (Taniguchi et al., 2001; OECD, 2003; Taniguchi & Thompson, 2015). The Physical Internet aims to transform independent freight systems into more open, collaborative, and integrated logistics networks (Montreuil et al., 2010; Montreuil, 2011). This involves shared use of resources and facilities, including freight vehicles and warehouses.

Digital twins provide an integrated suite of analytical, spatial mapping, and sensing procedures that can be used for promoting city logistics and the physical internet to improve the sustainability of urban freight systems. Digital twins provide a decision support system that interacts with the physical twin to improve the performance of complex and dynamic systems (NASEM, 2024). Digital twins consist of virtual data, analytics, and models that can be used to create information and aid decision-making for managing urban freight systems. Data collected is used to characterize the elements and the status of the physical twin.

Digital twins can be developed to improve operational decisions at freight terminals and transfer points in urban areas. It is important to minimize the waiting and search time of vans and trucks using loading zones and bays in central city areas. To promote low emission modes such as rail and zero emission vehicles such as electric vans and bikes, it is necessary to reduce delays and operational costs when transferring goods at terminals. Synchronization requires dynamic procedures for efficiently allocating resources at terminals and transfer points.

This chapter describes how digital twins can be developed and implemented at intermodal terminals, micro-consolidation centers, and physical internet hubs in urban areas. Applications of digital twins in the management of loading zones are also described. In addition, an overview of how spatial analysis tools can be utilized to aim the development and effectiveness of digital twins for hyperconnected city logistics is presented. Finally, a discussion of future directions is also provided.

12.2 Intermodal Terminals

The globalization of the world economy has significantly increased freight movement, with freight transport now vital to global supply chains. Despite its importance, the sector faces rising costs and inefficiencies, adding pressure on manufacturers (Capineri & Leinbach, 2006). In such an economic environment, the efficiency of intermodal terminals is crucial. In intermodal terminals, different modes of transportation—such as rail, road, sea, and air—converge to facilitate the transfer of cargo. Rail-road terminals include container terminals, equipped with cranes for moving containers between rail and road (Schulz et al., 2021). Sea-rail terminals, often located at ports, manage the transfer of containers between ships and rail systems, while ship-to-rail terminals handle direct cargo transfers from ships to trains (Yang et al., 2023). Sea-road terminals encompass roll-on/roll-off terminals for wheeled cargo and ferry terminals for vehicles and passengers. Air-rail terminals facilitate cargo movement between air and rail transport, with air cargo terminals located at airports and air-rail transfer facilities combining both modes. Air-road terminals, including airport cargo facilities and express delivery terminals, support the transition of cargo between air transport and road networks (Chen & Lin, 2016). Overall, intermodal terminals integrate various transport modes within a single facility for maximum flexibility, while specialized terminals, such as bulk cargo terminals and logistics hubs, are optimized for specific types of cargo and logistical operations.

Intermodal terminals face several significant challenges due to the complexities of modern freight operations. As containerized international trade has surged since the 1980s, the demand for efficient handling and storage in these terminals has increased, leading to issues such as inadequate space for container storage and inefficient utilization of loading equipment like cranes, automated guided vehicles, and straddle carriers (Jachimowski, 2017). The effectiveness of terminal technology and operations is heavily influenced by the volume of container flow, necessitating careful classification and organization based on terminal size and equipment. Problems such as resource allocation and scheduling of loading and unloading operations further compound these issues (Lin & Lin, 2016).

The lack of seamlessness in freight operations exacerbates these challenges, as inefficiencies in handling and processing can lead to increased costs and delays. Many intermodal terminals still operate with minimal support from information technologies, relying heavily on traditional policies developed from managerial experience. These policies are often adequate given current resources, but the rapid growth in freight transport demands more advanced infrastructure and management tools.

However, advancements in intramodality, technology, and logistics are addressing these issues by improving operational efficiency and reducing overall costs. For instance, innovative mathematical models and optimization techniques for resource allocation and scheduling have demonstrated potential for significant cost reductions and operational improvements. Simulation results confirm that optimized scheduling and resource management can reduce crane conflicts and truck queue lengths, highlighting the critical role of information flow in enhancing terminal performance. Effective information management is essential for overcoming these challenges and ensuring the smooth operation of intermodal terminals in the face of growing demand.

Various initiatives and methodologies have been suggested to improve the efficiency of intermodal terminals. For example, the European Intermodal Association has set minimum requirements for future intermodal terminals to ensure sufficient traffic concentration and economic management (ASCAME, 2024). However, before investing in new equipment or expanding terminal areas, operators are exploring whether new management methodologies can enhance performance. Computer-based simulations are proving valuable in this process, particularly for rail/road intermodal terminals, which differ from maritime terminals in scale and operational dynamics. These simulations help in modeling the complex interactions between rail and road networks, as well as the terminal's internal processes, thereby assisting in the development of effective strategies for terminal improvement.

12.2.1 Digital Twins for Intermodal Terminals

One solution to address the problems faced by intermodal terminals is to develop a digital twin (DT) for them. DT is an advanced technology that creates a virtual representation of a physical object, process, or system, integrating real-time data from its physical counterpart. This cyber-physical integration allows for continuous monitoring, analysis, and optimization throughout the lifecycle of the object or system. Key characteristics of DTs include real-time data integration, where data from sensors and other sources is used to dynamically update and accurately reflect the physical entity's status (Tao et al., 2019). Additionally, DTs employ advanced analytics to process large volumes of information, leading to enhanced predictions, decision-making, and performance optimization. They also support lifecycle management by providing insights from design and production through to maintenance and end-of-life. Furthermore, DTs enable real-time simulation and multi-disciplinary analysis, advancing from earlier device-specific and standard simulation approaches. Overall, DTs play a crucial role in smart manufacturing and operational intelligence, improving decision-making and system performance (Tao et al., 2022).

As an application of DT technology in terminals, a DT approach is used to design, develop, and implement high-precision autonomous docking control systems for container trucks by simulating dynamic behavior, testing, and optimizing control strategies before actual deployment (Widyotriatmo et al., 2024). Another example of DT technology in terminals is its use to optimize wireless communication systems. By creating virtual replicas of physical entities, the DT approach enables the application of techniques like Amplify-and-Forward (AF) to enhance terminal performance, resulting in a 17% improvement in Bit-Error-Rate and better signal quality (Dangana et al., 2024). Gao et al. (2024) use the DT technology in automated container terminals to enhance the efficiency of AGV scheduling by mapping the physical operation space into a synchronized virtual environment. This approach employs a mathematical programming model and Q-learning algorithm to address factors such as conflicts, failures, and battery constraints, demonstrating superior performance compared to genetic algorithms (GA) and particle swarm optimization (PSO) in both small- and large-scale scenarios.

The integration of Digital Twin (DT) technology in intermodal terminals offers a promising solution to address several operational challenges. Key issues such as coordinating vehicle arrivals and departures, managing queuing and delays, and addressing problems in booking systems—including no-shows and late arrivals—can be effectively managed through the live monitoring of terminal operations and real-time tracking of trucks and other terminal assets. Accurate prediction of arrival times is crucial for the optimal utilization of storage space and loading/unloading bays. Additionally, challenges related to the allocation of vehicle arrivals to specific bays and holding vehicles at gates due to delays significantly impact operational efficiency.

To enhance intermodal terminals' operations efficiency, there is a need for advanced models and real-time monitoring systems. The absence of platforms for managing transfers at hubs, inadequate real-time monitoring of bay and storage area utilization, and a lack of precise ETA prediction and gate allocation models hinder operational effectiveness. Improving vehicle consolidation and expediting last-mile deliveries require better synchronization of goods transfers between vehicles. The deployment of sensors to facilitate dynamic vehicle allocation and load management is essential. By incorporating DT technology, these challenges can be addressed through a comprehensive framework for real-time data integration and dynamic resource allocation, thereby improving operational coordination, reducing delays, and enhancing the ability to respond to disruptions while optimizing the use of available infrastructure.

Overall, a DT functions as an integrated system that combines multi-physics, multiscale, and probabilistic simulations to create a comprehensive representation of both the physical and virtual aspects of intermodal terminals. By incorporating physical products, virtual models, data streams, services, and the connections between them, a DT enables real-time monitoring and analysis of operational processes. This integration allows for precise tracking of terminal activities, optimization of resource allocation, and identification of inefficiencies. Furthermore, it provides a dynamic and detailed simulation environment where different scenarios can be tested and optimized before implementation, leading to improved decision-making and operational efficiency. Consequently, the use of a DT in intermodal terminals not only enhances the management of complex logistics but also supports proactive maintenance, strategic planning, and overall performance improvement.

12.2.2 DT's Application in Rail Freight Industry

Rail freight companies operate continuously, providing essential services throughout their extensive network of facilities. It is necessary. The seamless transportation of goods across mainland Australia is necessary to serve a diverse customer base across various industries. Rail companies move a broad mix of general freight cargo, import/export goods, and bulk commodities, including grain, coal, and minerals. Their commitment to safety, sustainability, and customer service underscores their pivotal role in maintaining and enhancing Australia's economic infrastructure. The operational philosophy of rail freight companies needs to focus on environmental, social, and governance (ESG). A commitment to long-term sustainability, emphasizing continuous improvement and well-being in the workplace, is also important. Rail companies can have terminals in all large cities of Australia, especially in Melbourne, which is the focus of this section.

Rail-road terminals in large cities typically encompass rail-mounted gantries (RMGs), reach stackers (RSs), rail tracks, a storage area for containers, and parking areas for trucks. Due to the diverse nature of freight operations and the variety of customers served, the operational procedures at the terminal can be inefficient.

Given the increasing demand for freight movement, this inefficiency could lead to suboptimal utilization of terminal capacity and potential financial losses for rail companies.

A primary cause of this inefficiency is the lack of adequate monitoring and measurement within the terminal. This issue can be mitigated by integrating a network of sensors and video cameras. These technologies will enhance the overall operational understanding of the terminal, providing additional data layers for various modeling purposes, such as image processing to detect movements within the terminal, and analytics to calculate the precise operational times of terminal machinery. Furthermore, implementing an integrated booking system, particularly for trucks, is essential to address this inefficiency. The arrival times of trucks can be affected by external factors, such as traffic congestion, which can negatively impact the scheduled plans for trains.

A DT is currently being developed for a major freight terminal in Melbourne, Australia. It integrates all modeling and analytics, thereby offering opportunities to evaluate operational decisions within a simulation-based environment. This simulation analysis allows for the examination of all feasible scenarios in the terminal. The development of such a DT, however, necessitates identifying all stakeholders involved in the terminal. Therefore, a stakeholder analysis was conducted to identify the major players in the terminal. These identified stakeholders are presented in Table 12.1.

The roles of stakeholders are crucial to identify, as their interactions and individual performance can significantly impact the terminal's operational efficiency. For instance, the terminal's capacity is a critical factor in determining how many customers can be served. There are various definitions of capacity for a rail-road terminal. Theoretical capacity is defined as "the capacity which the technical designers had in mind when commissioning the operation" (Slack et al., 2013), while nominal capacity is "the capacity that can be normally used, according to the number of production units, number of shifts per day, number of hours per shift, and number of days in each period" (Jonsson, 2008).

At the terminal, three different capacities are calculated and used for planning and operational procedures: storage capacity, referring to the maximum TEU[1] (twenty-foot equivalent unit) that can be stored on the ground in the terminal per year; equipment capacity, defined as the maximum TEU that the terminals equipment, including RMGs and reach stackers, can handle in a year; and track capacity, which is the maximum capacity of all terminal tracks per year. The overall capacity of the terminal is determined as the minimum of these three capacities. In this terminal, the bottleneck is equipment capacity, highlighting the importance of improving operational efficiency through DT technology. A DT can integrate all operational processes, from the booking system and truck entry to the final pick-up or drop-off of containers and exiting the terminal. This comprehensive integration

[1] TEU, or Twenty-Foot Equivalent Unit, is a standard measurement used in terminal operations to quantify cargo capacity, representing the volume of a standard 20-foot-long shipping container.

Table 12.1 Major stakeholders at a rail-road freight terminal

Stakeholders	Roles and objectives
Network managers	• To develop and maintain a high-quality rail network connecting terminals to required areas for fast and reliable goods movement, satisfying shippers. • To expand intermodal network infrastructure to meet increasing demand and align with government transport policy objectives.
Shippers	• To enhance reliability, flexibility, predictability, shorter transportation time, cost efficiency, and careful handling. • To minimize transportation and overall costs (with shippers prioritizing price sensitivity).
Terminal operator	• To enhance efficiency and reduce costs through faster transport to/from the terminal, streamlined customs clearance, value-added services, and simplified documentation. • To ensure reliable train connections and sufficient terminal capacity to meet shippers' logistical needs. • To monitor performance through key performance indicators.
Carriers	• To reduce the transportation and overall cost (carriers prioritize price sensitivity). • To minimize transport costs, including both out-of-pocket expenses and the value of transportation time.
Government	• To improve existing infrastructures and acquire new ones to enhance efficiency and accommodate growth. • To reduce the number of large truck movements in local communities surrounding terminal facilities.
Truck drivers	• To receive accurate information about shipment arrivals. • To receive streamlined container handling processes in terminals (implement ICT applications to facilitate straightforward and prompt handling of containers, reducing unnecessary delays for truck drivers).
Researchers	• To provide insights into other stakeholders, particularly terminal operators and the government, by modelling different processes of the terminal. • To investigate the potential impact of new-generation terminals on the competitiveness of intermodal transport.
Customers	• To have efficient and timely product delivery (by ensuring seamless coordination among stakeholders). • To optimize service delivery, including high-quality services with an appropriate cost.

enables a seamless flow of operations, optimizing resource utilization and enhancing the terminal's overall efficiency.

12.3 Micro-Consolidation Centers

Micro-consolidation centers (MCCs) are small logistics hubs typically located in inner urban areas. Goods are transported from large warehouses or hubs in outer urban areas to MCCs, where they are sorted, consolidated, and delivered to final customers (Kahalimoghadam et al., 2024a). MCCs are initiatives aimed at increasing the sustainability of freight systems and addressing unsustainable trends, such as rapid urbanization and pandemics (Kahalimoghadam et al., 2024c). A quantitative methodology was developed to select this alternative among other initiatives by various stakeholders (Kahalimoghadam et al., 2022), and the impacts of MCCs were evaluated in the Sydney metropolitan area to address the surge in e-commerce demands during the COVID-19 pandemic (Kahalimoghadam et al., 2021; Thompson et al., 2024).

Two challenges associated with MCCs are determining their optimal locations based on the geographical features of urban areas (Kahalimoghadam et al., 2024a) and developing a collaborative vehicle routing problem that minimizes the total vehicle kilometers traveled, including movements between MCCs and from MCCs to customers (Kahalimoghadam et al., 2024b). To address these challenges and facilitate the use of MCCs by various stakeholders, such as carriers and shippers, a DT can be a viable solution. A DT can be developed for a network of MCCs, where vehicle and goods movements are predicted, tracked, and monitored. Additionally, the DT can serve as a decision support system, aiding authorities in policy planning and strategic, tactical, and operational decisions.

12.4 Physical Internet Hubs

The Physical Internet (PI) framework offers a complementary approach to the integration of DT technology within the logistics domain. By emphasizing collaborative infrastructure and standardized goods transportation, the PI aims to optimize logistics operations and enhance the efficiency of logistical assets, such as containers, vehicles, and distribution centers (Montreuil et al., 2010). This aligns with the core principles of DT, which involve creating virtual representations of physical entities to improve decision-making and operational performance. PI-hubs function as compact consolidation centers strategically located within urban environments. These facilities foster collaboration among stakeholders to optimize delivery operations. By concentrating goods and shipments, PI hubs significantly reduce vehicle kilometers travelled and enhance the efficiency of the distribution network through increased consolidation rates. Furthermore, their proximity to end consumers enables expedited delivery services, making them particularly well-suited to the demands of the e-commerce industry.

An example of PI-hubs is the PI railroad terminal, introduced by Ballot et al. (2012), which was designed to address the drawbacks of traditional freight terminals, including emissions and congestion. Integrating the PI concept into a railroad terminal offers numerous benefits, including enhanced efficiency through standardized processes, real-time shipment visibility, and improved sustainability by optimizing resource use and reducing carbon emissions. This approach also supports scalability, cost reduction, and improved customer satisfaction through reliable and transparent services.

However, implementing the Physical Internet (PI) in multimodal terminals must consider the unique constraints of trains. These terminals typically serve a single line with limited stops, necessitating complex transportation schemes with other modes. Additionally, loading trailers onto trains can present gauge issues, and mismatched sizes between containers and trailers can compromise efficiency. Addressing these challenges is essential for the successful adoption of PI in railroad terminals. Digital Twin (DT) technology is well-suited for this purpose, as it provides a platform for seamless modeling and analytics, real-time data management, and monitoring the current state of all machines and assets in the terminal.

12.5 Loading Zones

Loading zones, also known as loading bays, are designated parking spaces for commercial vehicles to load and unload goods. They are extensively deployed in urban areas of large cities in Europe (Kalahasthi et al., 2022; Castrellon et al., 2024), Australia (Zhang et al., 2023), the USA (Department of Finance, 2023), and Mexico (Ochoa-Olán et al., 2021). These zones are important components of urban infrastructure, facilitating delivery activities. Compared to off-street loading docks, on-street loading zones are more desirable to couriers delivering parcels and small goods due to their easy accessibility. Despite their important role in urban freight systems, the supply of loading zones is often insufficient, which is due to the removal of on-street parking spaces aimed at improving traffic congestion and encouraging the use of public transport, cycling, and walking. Empirical studies have found that courier vehicles spend significant time (approximately 28% of the total trip time) on cruising for parking spaces due to high demand for loading zones, which substantially affects their delivery efficiencies (Chiara & Goodchild, 2020).

Understanding the utilization of loading zones and the parking behavior of courier vehicles is important for local governments to develop appropriate policies to balance the supply and demand for loading spaces. This includes determining the ideal number of loading zones and positioning them in high-demand areas, as well as implementing policies for loading zone management. Such policies may involve publishing historical occupancy or real-time availability information for loading zones (Chiara et al., 2022; Zhang & Thompson, 2019), dynamic parking pricing (Zheng & Geroliminis, 2016), and loading zone reservation systems (Mcleod & Cherrett, 2011; Comi et al., 2017).

Earlier studies of loading zone usage relied on survey data (Holguin-Veras et al., 2017; Thompson & Flores, 2016; Kim et al., 2021), which is not sustainable or feasible for long-term data collection over large regions. This limitation confines advanced methods for loading zone analysis and management to theoretical or simulation models or trial stages. However, with the development of sensing and detection technology, automated data collection from in-ground sensors (Comi et

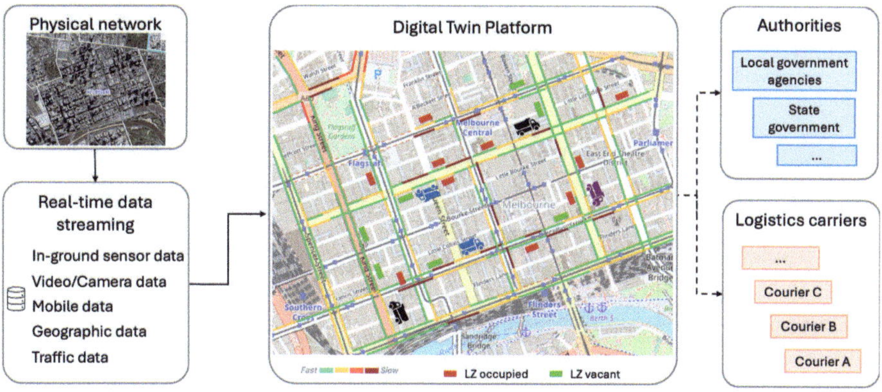

Fig. 12.1 Concept of a digital twin platform and its interaction with the physical network and selected urban freight stakeholders in the absence of any advanced analytical tools or methods. (Source of the aerial image of the City of Melbourne: https://maps.melbourne.vic.gov.au/)

al., 2018), mobile (GPS) devices (Diana et al., 2020), surveillance cameras (Elomiya et al., 2023), and other sources provides researchers and practitioners with access to large amounts of accurate and even real-time data. Consequently, this has stimulated numerous data-driven studies on loading zone-related logistics problems (Castrellon et al., 2024; Ludowieg et al., 2023).

Automated, real-time data transmission offers the opportunity to create digital representations of loading zones and integrate them into a unified digital twin (DT) platform of the urban freight system for monitoring, analyzing, and optimizing logistics activities. Besides providing real-time monitoring of loading zone availability and parking events, the digital twin platform enables the synthesis of loading zone data, traffic data, and vehicle trajectory data from various sources. Nevertheless, digital twins are merely virtual representations or visualizations of physical objects or systems if not integrated with analytical tools for understanding current and future events or decision-support systems for assisting decision-making in various circumstances. Figure 12.1 illustrates the concept of a digital twin platform for the urban freight network, focusing on loading zone-related activities, without any analytical tools. With the availability of real-time data streams, the platform can visualize the real-time usage of loading zones, traffic conditions, and track delivery vehicles as well as parcels/goods. Local governments can monitor the loading zones in real time, capturing illegal parking, such as double parking, overstaying, non-courier vehicle parking, and thereby enforcing parking regulations. Courier vehicles can utilize real-time information to plan their delivery routes, lowering the uncertainty in delivery to some extent. In addition, real-time tracking can improve customer service and ensure timely deliveries. However, without the support from analytical tools, the value that digital twins directly offer is significantly diminished. Individual stakeholders could stream data from the digital twin platform into their own decision-making processes, but this would limit the

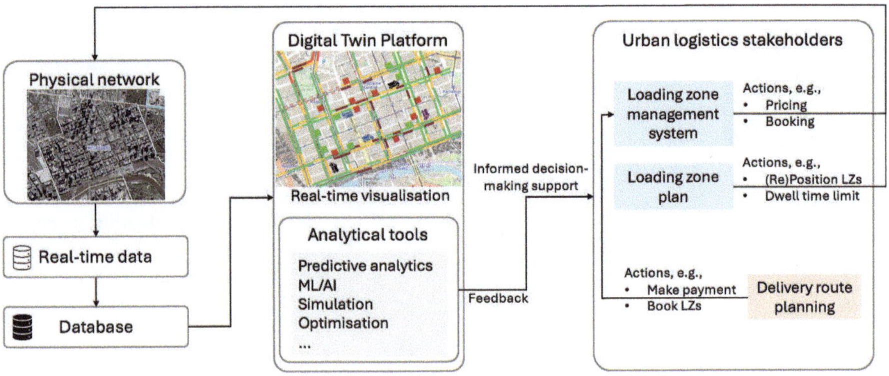

Fig. 12.2 Concept of interactions among the physical urban freight network, a digital twin platform, and loading zone-related urban freight stakeholders in the presence of advanced analytical tools. (Source of the aerial image of the City of Melbourne: https://maps.melbourne.vic.gov.au/)

opportunities for data sharing and collaboration, potentially leading to a waste of computational and physical resources. Furthermore, stakeholders with no or limited access to decision support tools cannot benefit much from the digital twin platform.

The synergy of data, predictive analytics, machine learning (ML), artificial intelligence (AI), simulation, and optimization techniques within a single digital twin platform can offer numerous benefits. These include enhancing the transparency and connectivity of the urban freight system, improving decision-making for stakeholders such as local authorities, policymakers, and courier companies, facilitating new initiatives or management systems, and ultimately improving efficiency, resilience, and sustainability of the urban freight system. Figure 12.2 presents the concept of integrating analytical tools into the digital twin platform and illustrates some interactions between the virtual representations and the physical users.

The digital twin platform can provide a transparent urban freight system via granting stakeholders access to real-time and historical data, thus addressing obstacles in data sharing and fragmentation in existing urban freight systems. Empowered by data analytics and statistical/ML techniques, the platform can help the local authorities understand the loading zone demand and utilization, as well as the relationship between land use (or the built environment) and freight demand (Wang et al., 2018). Furthermore, simulation models and predictive analytics can be applied to forecast long-term loading zone demand, informing the deployment of loading zones, specifically the number of loading zones, their layout and placement, and dwell time limits (Pinto et al., 2018; Castrellon et al., 2024). On the one hand, these prediction results can aid in planning infrastructure development and resource allocation. On the other hand, the prediction methods could benefit from additional information on existing land use and city development plans for forecasting future freight demand. Moreover, analyzing mobile data from courier vehicles can reveal new trends in delivery modes and patterns. For instance, the prevalence of walking as a delivery method is increasing in central city areas (Thompson et al., 2018). This insight can inform local government decisions on urban planning to facilitate multi-modal delivery options, such as using vans, electric vehicles, bikes, drones, and walking. Additionally, it can assist courier companies in making informed decisions on fleet sizing and personnel allocation, and route planning.

Besides supporting strategic decision-making, the benefits of the digital twin platform can be mostly realized at the operational level. By applying predictive analytics and simulation models, the platform can provide real-time predictions of freight vehicle demand, potential disruptions or risks, loading zone availability or occupancy, and traffic congestion levels (Ludowieg et al., 2023). Furthermore, with the assistance of optimization models and algorithms, it can inform couriers in optimizing their delivery strategies and routes (Muriel et al., 2022). Simulation models are powerful tools that have been widely used in studying urban delivery-related problems and transport networks. They can reproduce detailed behaviors of physical processes or objects in the digital twin system, including time-dependent, heterogeneous traffic, illegal parking, courier cruising, and competition. They can also allow bidirectional interactions between urban freight stakeholders and other transport network users, as well as between the stakeholders and the infrastructure. Therefore, they are particularly useful for what–if analyses to evaluate infrastructural, managerial, and optional alternatives, and to assess the impacts of one stakeholder's decisions on the others.

Although parking information systems can help reduce courier vehicle cruising distances and improve operational efficiency, they cannot completely eliminate the uncertainty of finding parking spots. Parking (loading zone) reservation systems, however, can secure parking spaces for courier vehicles in advance and minimize delivery uncertainty. Integrating the reservation system into the urban freight digital twin platform can benefit from the real-time monitoring of loading zone utilization, updates on traffic conditions, and tracking of courier vehicles, which enables courier vehicles to make a reservation in real time. While travel time uncertainty cannot be completely eradicated, information on the traffic conditions and the current location of courier vehicles can assist the system in predicting the vehicle arrival and departure times, making necessary adjustments in the reservation schedule.

While parking reservation and management systems are effective tools for freight demand management, dynamic pricing can proactively guide couriers' choices of delivery and parking time and locations, flattening demand curves and homogenizing spatial utilization. Couriers could be "passively" connected and coordinated by the system and "collaboratively" utilize parking resources. The digital twin platform can support dynamic real-time pricing strategies using real-time information on traffic congestion, parking facility utilization, and vehicle trajectories. By leveraging the integration of simulation models and optimization techniques, smart and effective parking management policies can be derived with diverse aims, e.g., improving the overall delivery efficiency and reducing the total environmental or economic costs.

In addition to the common challenges confronted in implementing digital twins for any large complex networks or systems, such as real-time streaming of high-resolution data, data fusion from diverse sources, the computational intensity of the analytical tools for real-time large-scale implementation, data accuracy, and data privacy, developing an urban freight digital twin platform must address the challenge from the fragmented urban freight system and overcome the data protection of freight stakeholders.

12.6 Spatial Platforms

Spatial data and technologies play a vital role in creating a more accurate and realistic digital representation of the physical world and processes in the Hyper-connected City Logistics context. In urban and regional areas, the urban digital twin platforms use spatial data as a foundation (ANZLIC, 2019) to represent a range of data and processes from single buildings to complex systems (Batty, 2018). These foundational data include positioning (coordinate systems), elevation, and urban entities' boundaries (cadastre). As a result of having such a foundation in place, in the context of an urban freight system, a map-base system can be built to incorporate the infrastructure (transport networks, roads, and loading zones), and other land uses and land covers (water bodies, vegetation, residential, and commercial areas). On top of such a map-base, other information such as utilities, social and economic, and real-time data (CCTV, IoT sensor network, and live traffic feed) can be added at different levels of fidelity (Fig. 12.3). As such, spatially explicit platforms enable a deeper understanding of the relationships between various elements within the twinned environment (Sabri et al., 2023a).

Achieving high-fidelity in urban digital twins (UDTs), for example, would necessitate incorporating precise spatial data related to buildings, infrastructure, and natural features. This emphasis on accuracy highlights how important spatial data is for creating effective modeling and simulation in digital twins. Spatial analysis, modeling, and simulation have been implemented in different domains such as human mobility (Miotti & Jain, 2021), indoor and outdoor foot traffic and evacuation models (Azimi et al., 2018; Tashakkori et al., 2016), and accessibility analysis (Sun et al., 2023). Furthermore, such platforms provide a capability for integrating emerging technologies such as geospatial AI (GeoAI), which has been used for more effective analysis of physics-based models such as hydrology (Gonzales-Inca et al., 2022), real-time geospatial-enabled traffic monitoring (Power et al., 2021), and large high-resolution mobility analysis (Lee et al., 2022).

It is important to note that visualization is a key aspect of spatial digital twins, and spatial data plays a significant role in creating meaningful visual representations. 3D models, maps, and other visualizations based on spatial data make the information within the digital twin more accessible and understandable to users, supporting decision-making and analysis. Literature highlights virtual reality (VR), augmented reality (AR), and extended reality (XR) as valuable tools for visualizing digital

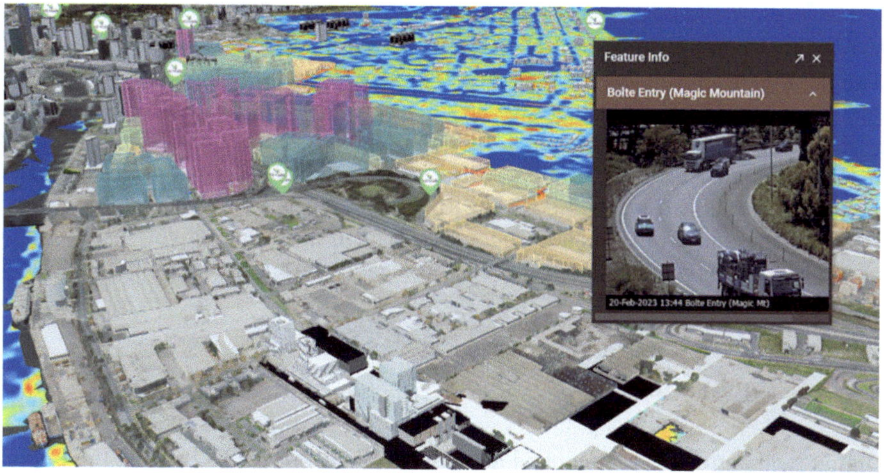

Fig. 12.3 Integrating multi-dimensional data (3D and real-time) in a spatially enabled Digital Twin, developed at the University of Melbourne for real-time impact analysis and water-sensitive design. Image: Centre for Spatial Data Infrastructure and Land Administration (CSDILA) Digital Twin team. Credit: University of Melbourne (from Sabri et al. (2023b), licensed under CC-BY 4.0)

twin data in a real-world context and collaborative planning, design, and decision-making (Stadtmann et al., 2023; Zhu et al., 2019). In particular, given the nature of hyperconnected city logistics involving multiple stakeholders and actors, XR plays an important role. According to Gong et al. (2021), XR is an umbrella term encompassing all computer-mediated reality technologies that blend the real and virtual realms, aiming to create an enriched user experience. XR encompasses a spectrum of technologies, from fully real to fully virtual environments, as represented by the reality–virtuality continuum, which can be used in different stages of design, development, and deployment of models in the urban freight digital twin.

To implement a spatial DT for urban freight systems, a cloud-edge-based architecture is necessary to integrate different layers of data, analytics, modeling and simulation, and visualization. In the hyperconnected city logistics, every physical object (built environment, vehicles, etc.), known as a "Thing," will be equipped with computing and communication features, enabling them to monitor the environment and collaborate through wired or wireless networks to deliver high-quality services to users. These progressively smarter connections and interactions, commonly referred to as Machine-to-Machine (M2M) communication or the Internet-of-things (IoT), integrated with the geospatial information (GeoIoT), play a key role (Alam & Saddik, 2017). Then these data can be integrated into multi-dimensional data (2D/3D) using geospatial standards (Web Services, CityGML, GeoJson) and connect to the application services. Notable service areas in a hyperconnected city logistics include Intelligent Transport Systems (ITS) and geospatial-enabled agent-based modeling and simulation.

However, it should be mentioned that existing urban digital twins fall short in leveraging geospatial data and analysis techniques, such as GeoAI and Geosimulation, which are crucial for real-time hyperconnected city logistics. To address this gap, a modular framework is necessary for encompassing data collection and integration, spatial analysis and modeling, and visualization using 3D models and game engines. By integrating real-time data, advanced spatial analysis, and immersive visualization, it is possible to enhance decision-making, improve risk management, and promote stakeholder engagement in collaboratively using urban capacities such as parking use and road optimization (Alexandridis et al., 2024).

12.7 Conclusions

Digital twins provide an integrated analytical framework for facilitating the implementation of hyperconnected city logistics that provides an effective approach for enhancing the sustainability of urban goods systems. Reducing the costs and delays at transfer points will lead to increased consolidation levels and less distance travelled by freight vehicles. Digital twins provide a practical means of coordinating operations at freight terminals and reducing delays at loading zones.

Adopting sustainable solutions in transportation and logistics is essential for mitigating negative externalities such as greenhouse gas and CO_2 emissions (Assarkhaniki et al., 2023). Smart electric vehicles (EVs), which are projected to reduce emissions by up to 43% compared to diesel engines, are pivotal in this transition (Bhatti et al., 2021). DT technology supports the mainstream adoption of EVs by providing detailed virtual models that optimize vehicle performance and sustainability. DT facilitates comprehensive simulation and analysis of various vehicle subsystems, including autonomous navigation, battery management, and vehicle condition monitoring. While DT offers significant advantages, including improved management of autonomous vehicles and reduced maintenance costs, several challenges must be addressed. These include data privacy and security concerns, the high costs associated with accurate modeling and the necessary network of sensors, and the overall expense of DT implementation.

References

Alam, K. M., & Saddik, A. E. (2017). C2PS: A digital twin architecture reference model for the cloud-based cyber-physical systems. *IEEE Access, 5*, 2050–2062.

Alexandridis, K., Zhang, S., Koohikamali, M., Sabri, S., & Ozkaya, E. (2024). Designing and implementing a robust, modular, and interoperable digital twin smart city framework for critical water spatial infrastructure. In *Proceedings of the 57th Hawaii International Conference on System Sciences (HICSS) 2024*. Scidler College of Business, University of Hawaii.

ANZLIC. (2019). *Principles for spatially enabled digital twins of the built and natural environment in Australia. The Australia and New Zealand Land Information Coun-*

cil (ANZLIC). https://www.anzlic.gov.au/resources/principles-spatially-enabled-digital-twins-built-and-natural-environment-australia

ASCAME. (2024). *European Intermodal Association*. https://www.ascame.org/

Assarkhaniki, Z., Sabri, S., Rajabifard, A., & Kahalimoghadam, M. (2023). Advancing sustainable development goals: Embedding resilience assessment. *Sustainability Science, 18*(5), 2405–2421.

Azimi, S., Delavar, M. R., & Rajabifard, A. (2018). Developing a multi-agent based modeling for smart search and rescue operation. *GeoInformation for Disaster Management Conference*, 133–157.

Ballot, E., Montreuil, B., & Thivierge, C. (2012). Georgia Southern commons functional design of physical internet facilities: A road-rail hub. *12th IMHRC Proceedings* (Gardanne, France – 2012).

Batty, M. (2018). Digital twins. *Environment and Planning B: Urban Analytics and City Science, 45*(5), 817–820.

Bhatti, G., Mohan, H., & Raja Singh, R. (2021). Towards the future of ssmart electric vehicles: Digital twin technology. *Renewable and Sustainable Energy Reviews, 141*(January), 110801.

Capineri, C., & Leinbach, T. R. (2006). Freight transport, seamlessness, and competitive advantage in the global economy. *European Journal of Transport and Infrastructure Research.*

Castrellon, J. P., Sanchez-Diaz, I., & Gil, J. (2024). Smart loading zones. A data analytics approach for loading zones network design. *Transportation Research Interdisciplinary Perspectives, 24*(2024), 101034.

Chen, X., & Lin, L. (2016). The integration of air and rail technologies: Shanghai's Hongqiao integrated transport hub. *Journal of Urban Technology, 23*(2), 23–46.

Chiara, G. D., & Goodchild, A. (2020). Do commercial vehicles cruise for parking? Empirical evidence from Seattle. *Transportation Policy, 97*, 26–36.

Chiara, G. D., Krutein, K. F., Ranjbari, A., & Goodchild, A. (2022). Providing curb availability information to delivery drivers reduces cruising for parking. *Scientific Reports, 12*, 19355.

City of Melbourne. (2018). *On-street parking data*. Available: https://www.melbourne.vic.gov.au/about-council/governance-transparency/open-data/Pages/on-street-parking-data.aspx. [Accessed: 04-Jun-2018].

Comi, A., Buttarazzi, B., Schiraldi, M. M., Innarella, R., Varisco, M., & Rosati, L. (2017). DynaLOAD: A simulation framework for planning, managing and controlling urban delivery bays. *Transportation Research Procedia, 22*, 335–344.

Comi, A., Schiraldi, M. M., & Buttarazzi, B. (2018). Smart urban freight transport: Tools for planning and optimising delivery operations. *Simulation Modelling Practice and Theory, 88*, 48–61.

Crainic, T. G., & Montreuil, B. (2016). Physical internet enabled hyperconnected city logistics. *Transportation Research Procedia, 12*, 383–398.

Dangana, M., Hussain, S., Ansari, S., Imran, M., & Zoha, A. (2024). A Digital Twin (DT) approach to Narrow-Band Internet of Things (NB-IoT) wireless communication optimization in an industrial scenario. *Internet of Things (Netherlands), 25*(February), 101113.

Department of Finance. (2023). *Commercial loading zone information*. https://www.chicago.gov/city/en/depts/fin/supp_info/revenue/CommercialLoadingZoneInfo.html. (Accessed March 2023).

Diana, M., Pirra, M., & Woodcock, A. (2020). Freight distribution in urban areas: A method to select the most important loading and unloading areas and a survey tool to investigate related demand patterns. *European Transport Research Review, 12.*

Elomiya, A., Křupka, J., & Jovčić, S. (2023). A smart parking system using surveillance cameras and fuzzy logic: A case study at Pardubice University's campus. *Procedia Computer Science, 225*(1).

Gao, Y., Chang, D., Chen, C. H., & Sha, M. (2024). A digital twin-based decision support approach for AGV scheduling. *Engineering Applications of Artificial Intelligence, 130*(December 2022), 107687.

Gong, L., Fast-Berglund, A., & Johansson, B. (2021). A framework for extended reality system development in manufacturing. *IEEE Access, 9*, 24796–24813.

Gonzales-Inca, C., Calle, M., Croghan, D., Torabi Haghighi, A., Marttila, H., Silander, J., & Alho, P. (2022). Geospatial Artificial Intelligence (GeoAI) in the integrated hydrological and fluvial systems modeling: Review of current applications and trends. *Water, 14*(14), 2211.

Holguin-Veras, C., Lawson, C., Wang, C., Jalccaller, M., Gonzalez-Calderon, C., Campbell, S., Kalahashti, L., Wojtowicz, J., & Ramirez-Rios, D. (2017). Using commodity flow survey microdata and other establishment data to estimate the generation of freight. *Freight Trips, and Service Trips: Guidebook NCFRP Research Report, 37.*

ITF. (2018). *Towards road freight decarbonisation – Trends, measures and policies, international transport forum*, Paris.

Jachimowski, R. (2017). Review of transport decision problems in the marine intermodal terminal. *Archives of Transport, 44*(4), 35–45.

Jonsson, P. (2008). *Logistics and supply chain management.* McGraw-Hill Higher Education.

Kahalimoghadam, M., Stokoe, M., Thompson, R. G., & Rajabifard, A. (2021). *The impact of COVID-19 pandemic on parcel delivery pattern in Sydney. 42nd Australasian Transport Research Forum (ATRF).*

Kahalimoghadam, M., Thompson, R. G., Rajabifard, A., & Stokoe, M. (2022). *Evaluation of governments' role in coping with city logistics problems. 43rd Australasian Transport Research Forum (ATRF), Paper 42.*

Kahalimoghadam, M., Thompson, R. G., & Rajabifard, A. (2024a). Determining the number and location of micro-consolidation centres as a solution to growing e-commerce demand. *Journal of Transport Geography, 117*(April), 103875.

Kahalimoghadam, M., Thompson, R. G., & Rajabifard, A. (2024b). Self-adaptive metaheuristic-based emissions reduction in a collaborative routing problem. *Sustainable Cities and Society, 110*, 105577.

Kahalimoghadam, M., Thompson, R. G., & Rajabifard, A. (2024c). Assessing unsustainable trends in city logistics. *Transportation Research Procedia, 79*, 401–408.

Kalahasthi, L. K., Sánchez-Díaz, I., Pablo Castrellon, J., Gil, J., Browne, M., Hayes, S., & Sentís Ros, C. (2022). Joint modeling of arrivals and parking durations for freight loading zones: Potential applications to improving urban logistics. *Transportation Research Part A, 166*, 307–329.

Kim, H., Goodchild, A., & Boyle, L. N. (2021). Empirical analysis of commercial vehicle dwell times around freight-attracting urban buildings in downtown Seattle. *Transportation Research Part A, 147*, 320–338.

Lee, A., Lee, K.-W., Kim, K.-H., & Shin, S.-W. (2022). A geospatial platform to manage large-scale individual mobility for an urban digital twin platform. *Remote Sensing, 14*(3), 723.

Lin, C.-C., & Lin, S.-W. (2016). Two-stage approach to the intermodal terminal location problem. *Computers and Operations Research, 67*, 113–119.

Ludowieg, A. R., Sanchez-Diaz, I., & Kalahasthi, L. K. (2023). Using machine learning to predict freight vehicles' demand for loading zones in Urban environments. *Transportation Research Record, 2677*(1), 829–842.

Mcleod, F., & Cherrett, T. (2011). Loading bay booking and control for urban freight. *International Journal of Logistics Research Applications, 5567*, 385–397.

Miotti, M., & Jain, R. (2021). Modeling aggregate human mobility patterns in cities based on the spatial distribution of local infrastructure. *Hawaii International Conference on System Sciences 2021 (HICSS-54).*

Montreuil, B. (2011). Toward a physical Internet: Meeting the global logistics sustainability grand challenge. *Logistics Research, 3*, 71–87.

Montreuil, B., Meller, R. D., & Ballot, E. (2010). Towards a physical Internet: The impact on logistics facilities and material handling systems design and innovation (2010). *11th IMHRC proceedings* (Milwaukee, Wisconsin. USA – 2010). 40.

Muriel, J. E., Zhang, L., Fransoo, J. C., & Perez-Franco, R. (2022). Assessing the impacts of last mile delivery strategies on delivery vehicles and traffic network performance. *Transportation Research Part C, 144*, 103915.

NASEM. (2024). *Foundational research gaps and future directions for digital twins, National Academies of Sciences, Engineering, and Medicine*. The National Academies Press.

Ochoa-Olán, J. J., Betanzo-Quezada, E., & Romero-Navarrete, A. J. (2021). A modeling and micro-simulation approach to estimate the location, number and size of loading/unloading bays: A case study in the city of Querétaro, Mexico. *Transportation Research Interdisciplinary Perspectives, 10*, 100400.

OECD. (2003). *Delivering the goods, 21st century challenges to Urban Goods Movement, Road Transport Research Programme (RTR), Directorate for Science, Technology, and Industry*. Organisation for Economic Development.

Pinto, R., Lagorio, A., & Golini, R. (2018). The location and sizing of urban freight loading/unloading lay-by areas. *International Journal of Production Research, 57*(1), 83–99.

Power, J., Jacoby, D., Sun, X., Plaudis, M., Drouin, M.-A., Coady, Y., & Meng, J. (2021). Real-time mission planning simulations from geospatial data. *2021 IEEE/ACM 25th international symposium on distributed simulation and real time applications (DS-RT)*, pp. 1–2.

Sabri, S., Alexandridis, K., Koohikamali, M., Zhang, S., & Ozkaya, E. (2023a, August). Designing a spatially-explicit urban digital twin framework for smart water infrastructure and flood management. *Proceedings of the 3rd Annual IEEE International Conference on Digital Twins and Parallel Intelligence (IEEE DTPI 2023)*, Orlando, FL.

Sabri, S., Winter, S., & Rajabifard, A. (2023b, March 2). *Creating digital twins to save our cities*. Pursuit. https://pursuit.unimelb.edu.au/articles/creating-digital-twins-to-save-our-cities

Schulz, A., Fliedner, M., Fiedrich, B., & Pfeiffer, C. (2021). Levelling crane workload in multi-yard rail-road container terminals. *European Journal of Operational Research, 293*(3), 941–954.

Slack, N., Brandon-Jones, A., & Johnston, R. (2013). *Operations Management* (7th ed.). Pearson.

Stadtmann, F., Rasheed, A., Kvamsdal, T., Johannessen, K. A., San, O., Kölle, K., Tande, J. O., Barstad, I., Benhamou, A., Brathaug, T., Christiansen, T., Firle, A.-L., Fjeldly, A., Frøyd, L., Gleim, A., Høiberget, A., Meissner, C., Nygård, G., Olsen, J., et al. (2023). Digital twins in wind energy: Emerging Technologies and industry-informed future directions. *IEEE Access*, 11, 110762–110795. IEEE Access.

Sun, Y., Olaru, D., Bright, C., McCarney, G., Reed, T. W., Sabri, S., Chen, Y., Amirebrahimi, S., Biermann, S., & Rajabifard, A. (2023). Making accessibility accessible: A flexible planning tool for enhanced urban analytics. *Research in Transportation Business and Management, 51*, 101042.

Taniguchi, E., & Thompson, R. G. (2015). In E. Taniguchi & R. G. Thompson (Eds.), *City logistics: Mapping the future*. CRC Press, Taylor & Francis.

Taniguchi, E., Thompson, R. G., Yamada, T., & van Duin, R. (2001). *City logistics – Network modelling and intelligent transport systems*. Elsevier.

Tao, F., Zhang, H., Liu, A., & Nee, A. Y. C. (2019). Digital twin in industry: State-of-the-art. *IEEE Transactions on Industrial Informatics, 15*(4), 2405–2415.

Tao, F., Xiao, B., Qi, Q., Cheng, J., & Ji, P. (2022). Digital twin modeling. *Journal of Manufacturing Systems, 64*(January), 372–389.

Tashakkori, H., Rajabifard, A., & Kalantari, M. (2016). Facilitating the 3D indoor search and rescue problem: An overview of the problem and an ant colony solution approach. *ISPRS Annals of Photogrammetry, Remote Sensing and Spatial Information Sciences, 4*(2).

Thompson, R., & Flores, G. (2016). Understanding deliveries to towers in Melbourne. *Transportation Research Procedia, 16*, 510–516.

Thompson, R. G., Zhang, L., Stokoe, M. (2018). Optimizing courier routes in Central Business Districts, *City Logistics 1*, Book Editors: Eiichi Taniguchi, Russell G. Thompson. Chapter 17, 325–341.

Thompson, R. G., Stokoe, M., Kahalimoghadam, M., & Mohri, S. (2024). ScienceDirect ScienceDirect Transforming distribution networks in metropolitan Sydney in response to COVID-19. *Transportation Research Procedia, 79*(2023), 154–161.

Wang, S., Zhang, L., & Thompson, R. G. (2018). A study of loading zone usage in Melbourne's Central City Areas. *The 24th international conference of Hong Kong society for transportation studies*, 8–10 December, 2018.

Widyotriatmo, A., Kuncara, I. A., Amri, H., Hasan, A., & Nazaruddin, Y. Y. (2024). Leveraging digital twin for autonomous docking of a container truck with stabilization control. *Journal of Field Robotics, 41*(3), 587–603.

Yang, Y., He, S., & Sun, S. (2023). Research on the cooperative scheduling of ARMGs and AGVs in a Sea–Rail automated container terminal under the Rail-in-Port Model. *Journal of Marine Science and Engineering, 11*(3).

Zhang, L., & Thompson, R. G. (2019). Understanding the benefits and limitations of occupancy information systems for couriers. *Transportation Research Part C, 105*, 520–535.

Zhang, L., Ding, P., & Thompson, R. G. (2023). A stochastic formulation of the two-echelon vehicle routing and loading bay reservation problem. *Transportation Research Part E, 177*, 103252.

Zheng, N., & Geroliminis, N. (2016). Modeling and optimization of multimodal urban networks with limited parking and dynamic pricing. *Transportation Research Part B, 83*, 36–58.

Zhu, Z., Liu, C., & Xu, X. (2019). Visualisation of the Digital Twin data in manufacturing by using Augmented Reality. *52nd CIRP conference on manufacturing systems.*

Part V
Advanced Integration and Future Directions

Chapter 13
Integrating AI and ML in Supply Chain Digital Twins: Bridging Potential and Foundational Research Gaps

Bulent Soykan and Ghaith Rabadi

Abstract Supply chains face increasing complexity, volatility, and demands for resilience, sustainability, and real-time responsiveness. These challenging conditions highlight the limitations of traditional optimization methods. The integration of artificial intelligence (AI) and machine learning (ML) with digital twins offers transformative potential for addressing these challenges through enhanced analysis, prediction, and automation. This chapter examines the critical synergy between AI/ML and digital twins specifically for supply chain optimization. It details how AI/ML capabilities can power intelligent data processing, advanced predictive analytics, predictive maintenance of assets, and pathways toward autonomous decision-making within supply chain digital twins (SCDTs). However, moving beyond conceptual potential, the chapter critically analyzes the significant foundational research gaps that currently impede the reliable, scalable, and trustworthy deployment of these integrated technologies. Key challenges synthesized include the pervasive need for robust and continual verification, validation, and uncertainty quantification (VVUQ) methodologies suited for dynamic AI/ML-driven systems, the development of theoretically sound and practical hybrid modeling approaches, scalable integration across complex systems of systems, managing heterogeneous and imperfect supply chain data, and addressing critical ethical, governance, and cybersecurity considerations. The chapter concludes that realizing the future vision of intelligent, optimized, and resilient supply chains is fundamentally contingent upon a concerted, collaborative, and interdisciplinary research effort focused on overcoming these mathematical, statistical, computational, and translational foundations, with a paramount focus on building trust and ensuring reliability.

B. Soykan (✉)
Institute for Simulation and Training, University of Central Florida, Orlando, FL, USA
e-mail: Bulent.Soykan@ucf.edu

G. Rabadi
School of Modeling, Simulation and Training, University of Central Florida, Orlando, FL, USA
e-mail: Ghaith.Rabadi@ucf.edu

G. Rabadi, B. Soykan (eds.), *Optimizing Supply Chains Through Digital Twins*,
International Series in Operations Research & Management Science 375,
https://doi.org/10.1007/978-3-032-08147-6_13

Keywords Supply Chain Digital Twins · Artificial Intelligence (AI) · Verification · Validation · and Uncertainty Quantification (VVUQ) · Hybrid Modeling · Trustworthy AI

13.1 Introduction

Contemporary supply chains operate within a dramatically altered global environment (Hugos, 2024). Extensive globalization has created highly interconnected and geographically dispersed networks (Ernst, 2002). While offering efficiency benefits, this global reach significantly increases inherent complexity and exposure to risk. Supply chains are now frequently subjected to unpredictable disruptions, market volatility, and geopolitical uncertainties (Manners-Bell, 2023). This environment places new and demanding pressures on supply chain operations. Businesses face an urgent need to build resilience, enabling them to anticipate, withstand, and recover quickly from shocks (Sheffi, 2015). Simultaneously, there is a growing imperative, driven by stakeholders and regulations, to enhance environmental and social sustainability across all supply chain activities (Chen & Kitsis, 2017). Also, competitive pressures and evolving customer expectations demand unprecedented levels of real-time visibility and responsiveness, which require agility to adapt quickly to changing conditions (Holloway, 2025). Together, these forces, globalization, volatility, and demands for resilience, sustainability, and real-time agility, characterize the complex and challenging landscape that confronts modern supply chains, necessitating advanced approaches to effective management and optimization.

Given the landscape described, traditional supply chain management methods often prove insufficient (Spekman et al., 1998). Many established approaches rely heavily on historical data analysis, periodic planning cycles, and relatively static optimization models. While effective in more stable environments, these methods struggle to cope effectively with the high degree of volatility, real-time demands, and intricate interdependencies inherent in modern supply chains. Plans based on past performance can quickly become outdated, and reactions based on delayed information can be too slow to mitigate disruptions or capture opportunities (Kleindorfer & Saad, 2005).

This inadequacy highlights a clear imperative for adopting more advanced optimization techniques. Central to this change is the need for real-time data. Accessing the latest information from across the supply network—spanning inventory levels, logistics movements, production status, and demand signals—provides the essential visibility required for timely and informed decision-making. Simply seeing the current state is not enough, however. There is also a critical need for predictive insights. Leveraging techniques that analyze current conditions and historical patterns allows organizations to better anticipate future events, such as potential bottlenecks, supplier risks, or fluctuations in customer demand, enabling proactive rather than reactive strategies (Aljohani, 2023).

The dynamic nature of the environment necessitates dynamic optimization. Supply chain plans and operational parameters must be capable of continuous adjustment in response to both real-time data feeds and predictive intelligence. This allows for ongoing optimization of resource allocation, logistics routing, production scheduling, and inventory policies to maintain efficiency and service levels amidst changing circumstances. Together, real-time data, predictive insights, and dynamic optimization represent important capabilities, driving the need for advanced analytical and technological solutions to effectively navigate modern supply chain complexities (Adewusi et al., 2024; Muthukalyani, 2024).

As a potential solution to navigate these complexities, the concept of digital twin is gaining significant attention within supply chain management and optimization. A digital twin represents a comprehensive virtual counterpart of a physical supply chain, including its assets, processes, and network interconnections (Biller et al., 2023; Soykan et al., 2025). Drawing from the formal definition proposed by the National Academies (NASEM, 2024), a supply chain digital twin (SCDT) can be understood as a collection of digital constructs that mimics the structure, context, and behavior of the actual supply chain system. Crucially, it is dynamically updated with real-time data streamed from its physical counterpart (e.g., sensors, logistics systems, enterprise resource planning systems).

Also, this virtual representation possesses predictive capabilities, allowing it to forecast future states or potential issues within the supply chain, and ultimately informs decisions designed to optimize performance and achieve business value. Central to this concept is the bidirectional interaction between the virtual and physical realms—data flows from the physical supply chain to continuously update the twin, and insights or automated commands flow back from the twin to influence or control physical operations. This dynamic connection and feedback loop fundamentally distinguish digital twins from traditional simulations or models (NASEM, 2024).

The potential of effectively implemented digital twins in supply chains is considerable. They promise enhanced end-to-end visibility, improved predictive forecasting of disruptions and demand, dynamic optimization of logistics and inventory, and the ability to test resilience strategies in a virtual environment (Khan, 2024). However, while enthusiasm for these digital twins is high, there are research gaps between the current widespread interest and aspiration surrounding the technology, and the established evidence base of proven, large-scale success (NASEM, 2024).

While the digital twin concept provides a powerful framework for representing and interacting with complex supply chains, realizing its full potential, particularly in dynamic and data-rich supply chain environments, depends critically on the integration of artificial intelligence (AI) and machine learning (ML). AI and ML are not merely enhancements; they serve as essential enabling technologies required to manage the inherent complexities and unlock the advanced capabilities envisioned for SCDTs.

The sheer volume, velocity, and a variety of data generated by and flowing into a comprehensive SCDT—from IoT sensors, logistics tracking, market feeds, and operational systems—quickly overwhelm traditional analytical methods. AI and ML techniques are uniquely suited for analysis at this scale, capable of sifting through vast datasets to identify subtle patterns, detect anomalies, and derive actionable insights that would otherwise remain hidden.

In addition, AI and ML significantly strengthen the predictive power of the digital twin. ML models can enhance forecasting accuracy for important supply chain variables like demand fluctuations, transportation times, potential disruptions, and equipment failures by learning from past data and real-time data streams. This moves the digital twin beyond simple mirroring toward providing valuable foresight.

The integration of AI and ML also paves the way for sophisticated automation within the supply chain, guided by the digital twin. Based on real-time analysis and predictions, AI-driven algorithms can enable automated decision-making for tasks such as dynamic route optimization, intelligent inventory adjustments, or proactive maintenance scheduling, driving operational efficiency and responsiveness. Therefore, AI and ML are fundamental components required to transform the SCDT from a digital model or digital shadow into an intelligent, adaptive, and value-generating system.

To provide a clear visual representation of how AI and ML integrate within the broader system, Fig. 13.1 illustrates the complete AI/ML-enhanced SCDT ecosystem. This diagram shows the cyclical flow, starting with real-time data from physical operations feeding the core digital twin model. It highlights the central role of the integrated AI/ML engine, performing the crucial functions of intelligent data processing, predictive analytics, and generating optimized decisions. These outputs then flow back to either inform human decision-makers or trigger automated actions within the physical supply chain, demonstrating the vital bidirectional interaction inherent in this approach.

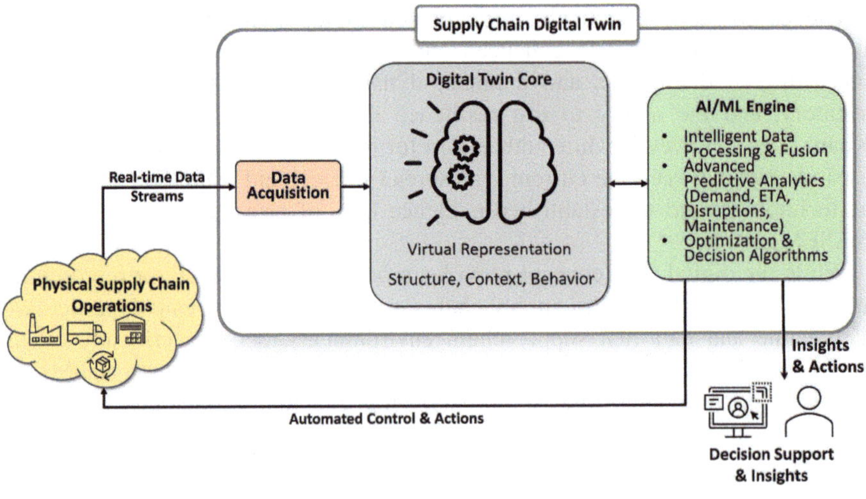

Fig. 13.1 The integrated AI/ML-enhanced supply chain digital twin ecosystem, showing the central bidirectional interaction

This chapter aims to provide a focused analysis of the synergy between AI, ML, and SCDT. The primary objective is twofold. Firstly, we will explore how AI and ML capabilities—specifically in areas such as advanced data analysis, predictive modeling, and automated decision-making—serve to significantly enhance the functionality and potential impact of digital twins applied to complex supply chain systems. Secondly, and critically, this chapter moves beyond simply outlining potential benefits. It aims to conduct a thorough examination of the foundational research gaps and future directions that currently impede the widespread, reliable, and effective deployment of these integrated technologies. This chapter will also identify and analyze key challenges in areas such as verification and validation, hybrid modeling, data integration, uncertainty quantification, ethical considerations, and system scalability. Therefore, the scope of this chapter encompasses not only the potential enhancements offered by AI and ML integration for SCDTs but also a critical assessment of the underlying scientific and technical challenges that must be overcome to realize this potential responsibly and effectively. The focus remains on the integration of AI/ML within the SCDT framework for optimization, contextualized by the necessary foundational advancements required for future success.

The remainder of this chapter is organized as follows: Sect. 13.2 provides the specific ways AI and ML enhance digital twins for supply chain management and optimization. It explores capabilities in real-time data analysis, predictive analytics, predictive maintenance, and autonomous decision-making, while also introducing associated research gaps. Section 13.3 provides a discussion centered on the foundational challenges and future directions essential for realizing trustworthy AI/ML-enhanced SCDTs. It synthesizes key research gaps, covering verification, validation, and uncertainty quantification (VVUQ), modeling, data, ethics, and implementation hurdles, and outlines the path forward. Section 13.4 concludes the chapter by summarizing the key insights regarding the synergy of AI, ML, and SCDTs, reiterating the critical importance of addressing the identified foundational challenges to unlock the technology's full potential.

13.2 Enhancing Supply Chain Digital Twins with AI and ML: Capabilities and Gaps

13.2.1 AI for Intelligent Data Processing and Real-Time Analysis in Supply Chains

The effective operation of an SCDT is fundamentally predicated on its ability to ingest and interpret vast quantities of diverse, rapidly changing data (Maheshwari et al., 2023). Modern supply chains generate an unprecedented volume of data, from granular sensor readings on manufacturing equipment and transported goods to large-scale transactional records (Neway, 2024). Data also arrives at high velocity, requiring near-real-time processing of streaming information like GPS locations from logistics fleets or continuous updates from warehouse management systems. In addition, this data exhibits significant variety, encompassing structured numerical data (e.g., inventory counts, transit times), semi-structured data (e.g., EDI messages, log files), and unstructured data (e.g., supplier emails, weather reports, news feeds, images of cargo). AI and ML are indispensable for handling this complex data landscape. AI/ML algorithms excel at processing massive datasets, identifying patterns, detecting anomalies, and extracting meaningful insights far beyond the capabilities of traditional analytics or human oversight, enabling the SCDT to build an accurate and timely understanding of the supply chain's state (Rathore et al., 2021).

Despite AI's potential, significant foundational research gaps hinder its effective application for intelligent data processing within SCDTs. A primary challenge involves integrating multimodal and heterogeneous data streams specific to the supply chain domain. Supply chains inherently involve combining data with vastly different formats, semantics, quality levels, and timescales—for example, correlating quantitative sensor data from a refrigerated container with qualitative text-based customs reports or volatile market pricing data. Developing AI/ML techniques that can effectively fuse and reason across these disparate data types, potentially requiring advancements in areas like common data models, semantic interoperability frameworks, and sophisticated data fusion algorithms, remains an important research need.

Compounding this challenge is the critical issue of ensuring data quality and provenance for AI inputs. AI models are notoriously sensitive to data quality; inaccuracies, biases, or missing information in the input data can lead to flawed analyses and unreliable predictions from the digital twin. In supply chain, data often originates from numerous sources across organizational boundaries (suppliers, logistics providers, customers), making data quality control and verification difficult. Documenting data quality and provenance is important. There is a pressing need for AI/ML methods that can automatically assess the quality and reliability of incoming data streams or, alternatively, AI/ML algorithms that are inherently robust to the imperfect, noisy, and potentially incomplete data characteristic of real-world supply chain environments.

The demand for real-time responsiveness in the supply chain requires highly scalable AI/ML algorithms optimized for real-time analytics. Many advanced AI/ML algorithms require significant computational resources and time for both training and inference. For an SCDT guiding operational decisions, analysis must often occur in near real time to be actionable. This presents a significant gap in developing AI/ML algorithms that are not only accurate but also computationally efficient and scalable, capable of running continuously on potentially resource-constrained platforms (including edge devices deployed within the supply chain) to support the dynamic nature of the SCDT. Addressing these gaps related to data integration, quality assurance, and scalable real-time processing is fundamental to fully leveraging AI for intelligent data analysis within the next generation of SCDTs.

13.2.2 ML for Advanced Predictive Analytics in Supply Chains

A core promise of integrating ML into SCDTs lies in enhancing advanced predictive analytics (Singh, 2023). ML algorithms enable the digital twin to move beyond simply reflecting the current state or performing basic extrapolations, allowing it to learn complex patterns from large amounts of historical and real-time data to anticipate future events with greater accuracy. This predictive capability is crucial for proactive supply chain. For instance, in demand forecasting, ML models can analyze intricate relationships between sales data, seasonality, promotional activities, competitor actions, macroeconomic indicators, and even unstructured data like social media trends, often outperforming traditional statistical methods, especially for products with complex demand patterns or when forecasting across various levels of granularity (e.g., SKU, location, channel). Similarly, ML significantly improves ETA prediction in logistics by dynamically integrating real-time variables such as traffic conditions, weather forecasts, driver behavior, vessel telemetry, and port congestion data, leading to more accurate delivery timelines and better resource planning. In addition, ML can play a vital role in disruption prediction, identifying subtle leading indicators of potential problems—like degrading supplier performance metrics, specific geopolitical news patterns, or unusual weather phenomena upstream—that might signal future delays, shortages, or capacity constraints, enabling preemptive mitigation actions.

However, despite these promising applications, the reliable and widespread use of ML for predictive analytics within SCDTs is hampered by several significant foundational research gaps:

- **VVUQ for ML Predictions, Especially in Extrapolation and for Rare Supply Chain Events:** ML models often struggle when predicting outcomes under conditions not well-represented in their training data (extrapolation) or when forecasting rare but high-impact events (e.g., pandemics, major geopolitical crises, natural disasters severely impacting logistics hubs). Standard VVUQ techniques are often inadequate for these scenarios (Committee on Mathematical Foundations of Verification, Validation, and Uncertainty Quantification National Academies of Sciences, Engineering, and Medicine, 2012). There is a lack of standards for reporting ML model confidence and the critical need for continual VVUQ methods suited for dynamic systems and extrapolatory conditions. In supply chain, decisions based on unreliable extrapolations or failed predictions of rare events can lead to significant financial losses, operational paralysis, or failure to meet customer needs. There is a pressing need for research into robust VVUQ methodologies specifically designed for ML models operating in these challenging supply chain contexts, including better ways to quantify uncertainty when extrapolating.
- **Developing ML Models Robust to Sparse/Noisy Supply Chain Data:** Real-world supply chain data is frequently imperfect (Lechler et al., 2019). It can be sparse (e.g., limited sales history for new products, infrequent data points from certain logistics segments), noisy (e.g., sensor errors, manual data entry

mistakes, inaccurate timestamps), or incomplete. ML models trained primarily on clean, dense datasets may exhibit poor performance and lack robustness when deployed on typical supply chain data streams. This gap limits the practical applicability of many advanced ML techniques. Research is needed to develop ML architectures and training strategies inherently resilient to these data imperfections, or alternatively, more effective methods for data cleaning, imputation, and anomaly handling tailored to supply chain data characteristics.
- **High Cost of Training Data:** Acquiring, cleaning, labeling, and maintaining the large, high-quality datasets often required to train sophisticated ML models can be prohibitively expensive and time-consuming in a supply chain setting (Long et al., 2025; Khedr et al., 2024). The substantial costs associated with generating sufficient training data, especially data covering a wide range of operational scenarios and rare events, are often overlooked but represent a major barrier (Mihai et al., 2022). This limits the adoption of advanced predictive analytics, particularly for smaller enterprises or for highly specific supply chain applications. Addressing this requires advancements in data-efficient ML approaches suitable for supply chain, such as transfer learning, few-shot learning, self-supervised learning, and the development of high-fidelity synthetic data generation techniques.
- **Lack of Standards for Reporting ML Model Confidence:** A significant hurdle for using ML predictions in supply chain decision-making is the frequent lack of clear, reliable, and standardized measures of confidence associated with those predictions (Baryannis et al., 2019). Decision-makers need to understand the level of uncertainty in a forecast (e.g., demand prediction, estimated time of arrival (ETA) window) to assess risks and make informed choices. There is a clear lack of standards in reporting model confidence. Without such standards, interpreting the trustworthiness of an ML prediction becomes subjective, hindering its adoption for critical supply chain planning and execution tasks. Establishing and adopting standardized metrics and communication protocols for conveying the confidence and uncertainty of ML-based supply chain predictions is important for establishing trust and facilitating responsible use.

13.2.3 AI/ML-Driven Predictive Maintenance for Supply Chain Assets

Predictive maintenance (PdM) represents a significant value proposition for integrating AI and ML within digital twins. The goal is to shift from traditional reactive (repair after failure) or preventative (time-based) maintenance schedules toward a proactive, condition-based approach (Abd Wahab et al., 2024; Hegazy Abdelghany Mohamed Ammar, 2021). In the context of supply chain, this applies to a wide array of critical physical assets, including logistics equipment (trucks, ships, railcars, delivery drones, forklifts), warehouse automation systems (conveyor belts, robotic arms, automated guided vehicles), and manufacturing machinery involved in production nodes of the supply chain. Failures in any of these assets can cause significant disruptions, leading to delays, increased costs, safety risks, and damaged customer relationships. AI/ML algorithms, operating within the digital twin framework, are pivotal in enabling effective PdM by analyzing operational data to forecast potential failures before they occur.

The process typically involves leveraging data streams from sensors embedded within these assets—monitoring parameters like vibration, temperature, pressure, oil quality, acoustic signatures, power consumption, operational hours, GPS data, and error logs. AI/ML models are trained on this data, often combined with historical maintenance records, to learn complex patterns indicative of wear, degradation, or impending malfunction. Specific AI/ML tasks include:

- **Anomaly Detection:** Identifying subtle deviations from normal operating baselines that may signal the incipient stages of a fault, often earlier than traditional threshold-based alarms (Huang et al., 2021; Xu et al., 2021).
- **Failure Prediction/Classification:** Estimating the probability of specific failure modes occurring within a defined future timeframe, allowing maintenance teams to prioritize interventions (Alshathri et al., 2023).
- **Remaining Useful Life (RUL) Estimation:** Quantifying the expected operational lifespan left for a component or asset before maintenance is likely required, facilitating optimized scheduling and parts procurement (Kobayashi & Alam, 2024; Anis et al., 2020; He et al., 2021).

The SCDT acts as the integration platform, collecting the sensor data, hosting the AI/ML PdM models, potentially running simulations to test failure scenarios or maintenance strategies, and delivering actionable alerts and recommendations to maintenance planners and technicians. However, reliably implementing AI/ML-driven PdM within the dynamic and complex environment of supply chain faces substantial research gaps:

Applying PdM Methods Reliably in Dynamic Supply Chain Operating Conditions Unlike assets in highly controlled environments, supply chain equipment often operates under constantly dynamic and variable conditions (Sarimveis et al., 2008). Logistics vehicles encounter diverse terrains, weather patterns, traffic congestion, payload variations, and differing driver behaviors. Warehouse automation systems face fluctuating throughput demands and product mixes. Manufacturing equipment within supply nodes may be subject to changes in raw materials, production schedules, and operator skill levels. AI/ML models trained primarily on data from one set of operating conditions may fail to generalize reliably when deployed across the diverse and fluctuating realities of a global supply chain. This lack of robustness can lead to inaccurate predictions, false alarms, or missed failures. Addressing this gap requires developing AI/ML models that are inherently resilient to these operational variations or possess sophisticated adaptive learning capabilities. These models must be able to continuously update and recalibrate themselves within the SCDT framework as operating contexts shift, demanding advancements in online learning, transfer learning, and domain adaptation techniques specifically tailored for the nonstationary nature of supply chain operations. This challenge directly connects to the need for continual VVUQ, ensuring models remain trustworthy as conditions evolve.

VVUQ for Maintenance Predictions Providing a deterministic prediction (e.g., "Component X will fail in 7 days") is often insufficient and potentially misleading for critical supply chain maintenance decisions. Failures are stochastic events, and predictions are inherently uncertain due to data noise, model limitations, and unpredictable operating factors. Supply chain decision-makers require a clear understanding of the uncertainty associated with PdM forecasts to make risk-informed choices regarding maintenance scheduling, resource allocation, and operational contingency planning. Robust VVUQ is essential for building trust, yet methodologies for rigorously quantifying and communicating the uncertainty of AI/ML-based PdM predictions (especially for complex models like deep neural networks) are underdeveloped. There is a significant gap in developing scalable VVUQ techniques that can provide reliable confidence intervals for RUL estimates or calibrated probabilities for failure predictions, particularly when dealing with high-dimensional sensor data or predictions of rare but critical failure modes where historical failure data is scarce. In addition, establishing standardized metrics and reporting formats for this uncertainty is important for practical adoption by maintenance teams.

13.2.4 Toward Autonomous Optimization and Decision-Making in Supply Chains

Beyond enhancing predictive capabilities and maintenance strategies, a significant long-term potential for integrating AI and ML within SCDTs lies in enabling autonomous optimization and decision-making. This represents a shift from the digital twin merely providing insights or recommendations for human action toward a system where the digital twin, powered by AI/ML, can directly initiate and execute operational adjustments within predefined boundaries. The potential scope encompasses various tactical and operational decisions across the supply chain, including:

- **Autonomous Routing:** Dynamically rerouting logistics vehicles (trucks, ships, drones) in real time based on live traffic data, weather updates, delivery window changes, fuel optimization calculations, or predicted congestion, with instructions sent directly to navigation systems (Zhang et al., 2024; Lou et al., 2023)
- **Automated Inventory Management:** Triggering replenishment orders automatically when inventory levels, combined with ML-based demand forecasts and lead time predictions, fall below dynamically calculated reorder points; or

executing automated balancing transfers between distribution centers based on network-wide optimization (Maheshwari et al., 2023)

- **Intelligent Sourcing and Allocation:** Automatically allocating orders among preapproved suppliers based on real-time performance data, risk assessments derived from news feeds or financial indicators, predicted capacity, and dynamic cost structures, adhering to predefined business rules and contracts
- **Adaptive Process Control:** Making minor, automated adjustments to parameters in warehouse sorting systems or production line schedules based on real-time throughput monitoring and short-term demand forecasts generated by the SCDT

The underlying mechanism involves the SCDT continuously assimilating real-time data to maintain an accurate representation of the supply chain state. AI/ML algorithms analyze this state, run predictive models, and feed the results into optimization routines to determine the best course of action. In an autonomous setup, the system would then translate this optimal decision into executable commands for the relevant physical assets or enterprise systems (e.g., updating an order in the enterprise resource planning (ERP), sending new coordinates to a vehicle). However, achieving this vision of reliable and effective autonomous supply chain decision-making requires overcoming substantial foundational research gaps:

Scalable Optimization Under Uncertainty (Incorporating Risk Metrics and Extreme Events) Supply chain decisions are rarely made under certainty; demand, lead times, costs, and potential disruptions are all stochastic. Autonomous systems cannot simply optimize for the average or most likely scenario. They must robustly account for uncertainty and risk. There is a need for optimization methods that incorporate advanced risk metrics (going beyond simple probabilities) and can handle low-probability, high-impact extreme events (e.g., port shutdowns, supplier bankruptcies, natural disasters). In addition, solving these complex, uncertainty-aware optimization problems at the scale of a global supply chain, potentially in near real time, requires significant advancements in scalable algorithms. The current state of the art often struggles with the dimensionality and complexity required for truly autonomous supply chain optimization under realistic uncertainty.

Integrating AI/ML with Data Assimilation (DA) and Control Theory An autonomous system operates within a continuous feedback loop: sensing the environment (DA updates the SCDT), processing information (AI/ML analysis and prediction), deciding on an action (optimization/control), and affecting the environment (executing the decision). DA alone lacks the learning capability needed, and the synergy between models and data in decision-making is critical. There is a gap in integrating modern AI/ML techniques with established principles from DA and control theory within a unified framework. This includes developing methods where ML models inform or are part of the DA process, and where AI/ML (such as reinforcement learning (Soykan and Rabadi, 2024)) is integrated with control strategies for dynamic supply chain problems, all while managing the associated uncertainties and ensuring stability.

Developing Methods for AI-driven Sensor Steering/Optimal Experimental Design (OED) An intelligent autonomous supply chain system should actively manage its own data acquisition to improve decision quality. Optimally designing and steering data collection requires developing methods for AI-driven sensor steering and OED. For example, an autonomous system might determine that getting more frequent location updates from trucks in a specific weather-affected region is more valuable than broad, uniform polling, or that initiating a specific quality check sensor on a batch of components is warranted based on upstream process variations detected by the SCDT. Developing scalable AI/ML algorithms that can perform this goal-oriented optimization of data collection specifically to enhance autonomous decision-making in real time is a key research gap.

Need for Hybrid Modeling Approaches for Robust Decisions Purely data-driven AI/ML models can be brittle and unreliable when faced with novel situations or when needing to respect underlying physical or logical constraints common in supply chain (e.g., truck capacity limits, warehouse throughput physics, contractual obligations). There are limitations in current hybrid modeling approaches that combine physics-based/domain-knowledge models with data-driven ML. For autonomous systems, where incorrect or infeasible decisions can have significant consequences, ensuring robustness is important. There is a need for theoretically grounded hybrid modeling techniques that effectively integrate supply chain domain knowledge and constraints with the pattern-recognition strengths of ML, leading to more reliable, trustworthy, and interpretable decisions suitable for automation.

13.3 Foundational Challenges and Future Directions for AI/ML-Enhanced Supply Chain Digital Twins

While the integration of AI and ML holds immense promise for enhancing SCDTs, realizing this potential reliably and at scale is contingent upon addressing several profound, cross-cutting foundational research gaps. These gaps pertain not merely to specific applications or algorithms, but to the underlying mathematical, statistical, and computational principles required for trustworthy and effective cyber-physical systems like SCDTs. Several key areas emerge as particularly critical impediments demanding focused research:

Comprehensive and Continual VVUQ Traditional VVUQ often focuses on static models, but SCDTs, especially those driven by AI/ML, evolve over time as they assimilate new data and operating conditions change. There is a critical lack of established methods and standards for verifying and validating these evolving systems, particularly the AI/ML components which can act as "black boxes." Quantifying uncertainty is especially challenging when SCDTs are used for extrapolation beyond their training data or for predicting rare but critical supply chain events, scenarios common in risk management and resilience planning. Ensuring the trustworthiness of AI/ML-driven supply chain decisions absolutely depends on developing and implementing robust, computationally feasible, and continual VVUQ frameworks.

Theory and Methods for Hybrid Modeling There is a need for synergistic hybrid modeling approaches that effectively combine data-driven AI/ML techniques with mechanistic, physics-based, or domain-knowledge models. Purely data-driven models may fail to respect fundamental physical laws or operational constraints inherent in supply chains, while purely mechanistic models may struggle with complexity or unmodeled phenomena. Hybrid models offer a path to leverage the pattern-recognition strengths of AI/ML while grounding predictions in established domain principles, potentially improving robustness, interpretability, and data efficiency. Developing these hybrid approaches is critical for creating reliable AI/ML-driven SCDTs capable of robust decision-making in complex supply chain environments.

Integration and Scalability of Complex Systems SCDTs typically represent complex systems of systems, involving multiple interconnected components (e.g., manufacturing nodes, warehouses, logistics fleets, suppliers) potentially represented by different models or even distinct DTs. The integration of component/subsystem digital twins, especially those operating at different scales or fidelities, is a significant challenge. Ensuring stable, consistent, and verifiable behavior across these coupled systems requires foundational advancements in multimodel coupling techniques and compositional VVUQ. In addition, the scalability of algorithms (for data processing, AI/ML training/inference, optimization, VVUQ) and the supporting computational infrastructure is a major barrier, particularly given the demanding data volumes and computational intensity associated with sophisticated AI/ML models applied to large-scale supply chain problems.

Data Integration, Quality, and Standardization The effectiveness of any AI/ML-driven SCDT is fundamentally dependent on the quality and usability of its input data. The significant challenges are in integrating heterogeneous and multimodal data streams common in supply chain, stemming from a lack of adopted standards for data formats, metadata, and semantics, which hinders interoperability. Robustly handling data quality issues—including missing data, noise, bias, and ensuring trustworthy data provenance—is critical, particularly for training reliable AI/ML models. There is a research gap in developing scalable methods for data fusion, automated quality assessment, and frameworks for standardized data and metadata exchange suitable for the dynamic, cross-organizational nature of supply chains.

13.3.1 Implementation Challenges in the Supply Chain Context

The path toward widespread adoption of AI/ML-enhanced SCDTs includes considerable practical implementation challenges. These challenges often relate to the complexities of real-world operational environments, existing technological landscapes, organizational structures, and economic considerations. Addressing these practical barriers is as crucial as advancing the underlying science. Key implementation challenges include:

Integrating Disparate Supply Chain IT Systems Modern supply chains rely on a patchwork of specialized IT systems—ERP systems, warehouse management systems, transportation management systems, manufacturing execution systems, supplier relationship management portals, IoT platforms, and numerous legacy applications. These systems often originate from different vendors, utilize diverse data formats and communication protocols (APIs), and were frequently implemented in silos without interoperability as a primary design goal. Attempting to integrate these disparate systems to feed a comprehensive SCDT with consistent, timely, and accurate data is a challenging system integration challenge. It often requires substantial investment in middleware, custom interfaces, data cleansing and transformation processes, and ongoing maintenance, significantly increasing the complexity and cost of SCDT deployment (Themistocleous et al., 2004).

Data Sharing Across Organizational Boundaries A truly effective SCDT often requires data that extends beyond the confines of a single organization, incorporating information from suppliers, logistics providers, contract manufacturers, distributors, and potentially even end customers. However, facilitating secure and reliable data sharing across organizational boundaries presents major obstacles. Issues of trust, concerns regarding data security, protection of intellectual property and competitive sensitivities, the lack of standardized data sharing agreements and technical protocols, and conflicting data governance policies all create significant friction. Without robust mechanisms and agreements to overcome these barriers, the SCDT may remain fragmented, lacking the true end-to-end visibility needed to optimize the entire value chain.

Cybersecurity of Interconnected Supply Chains The very nature of a digital twin—creating a tight, bidirectional link between the physical supply chain and its virtual representation—inherently expands the cybersecurity attack surface. This interconnectedness, especially when spanning multiple organizations and involving control capabilities, introduces heightened risks. Malicious actors could potentially target the SCDT or its data feeds to glean sensitive operational intelligence, inject false data to mislead decision-making, manipulate the virtual model to cause physical disruptions (e.g., altering logistics routes, disrupting production parameters), or use a compromised partner's connection as a gateway into other systems. Securing this complex cyber-physical ecosystem requires a holistic approach encompassing not just the SCDT platform itself, but also the diverse data sources, communication networks, integrated IT/operational technology (OT) systems, and the physical assets being monitored and potentially controlled across the extended supply chain network.

Cost Justification vs. Hype Developing, implementing, and maintaining a sophisticated SCDT, particularly one leveraging advanced AI/ML, represents a significant financial and resource commitment. Quantifying the return on investment (ROI) can be challenging, as benefits like enhanced resilience, improved agility, or better risk management are often difficult to translate directly into near-term financial gains. The considerable hype surrounding digital twins and AI can lead to unrealistic expectations among stakeholders. Separating the aspirational potential from demonstrable, near-term value is crucial. Failure to deliver on inflated expectations, especially in early pilot projects, can lead to disillusionment and withdrawal of support. Building a compelling, phased business case that focuses on achievable milestones and clearly articulates tangible benefits—moving beyond buzzwords—is essential but often difficult for complex, cross-functional initiatives like SCDTs.

Table 13.1 provides an overview of these key foundational research gaps. The table categorizes the primary obstacle areas—ranging from VVUQ tailored for AI/ML systems, to hybrid modeling complexities, scalability issues, data integration hurdles, challenges in autonomous decision-making under uncertainty, and essential ethical, governance, and security considerations. Each listed gap represents a significant challenge, highlighting specific needs pertinent to leveraging AI and ML effectively within the demanding SCDT context. Addressing these interconnected gaps comprehensively, as summarized in the table, is deemed essential for advancing the field beyond its current state and realizing the full potential of the SCDTs for optimizing future supply chains.

13.3.2 Ethical, Governance, and Security Considerations

The deployment of AI/ML-enhanced digital twins within complex and interconnected supply chain ecosystems requires careful and proactive consideration of significant ethical, governance, and security challenges. These concerns move beyond purely technical feasibility and are crucial for building trust, ensuring responsible adoption, and mitigating potential negative consequences. Several key areas demand particular attention:

Data Privacy and Ownership Across Supply Chain Partners SCDTs often achieve maximum value when they integrate data streams originating from multiple independent organizations—suppliers, logistics providers, manufacturers, distributors, and potentially customers. In supply chains, challenges surrounding protecting privacy and determining data ownership and liability involve navigating the sharing of commercially sensitive information (e.g., real-time inventory levels, production costs, supplier performance metrics, negotiated pricing), potentially personally identifiable information (PII) linked to drivers or warehouse staff (e.g., location tracking, performance monitoring), and detailed operational data that could reveal competitive insights. Ambiguity over who owns the raw data versus the derived insights, how data usage is controlled and audited across partners, and how privacy is maintained for both individuals and organizations creates significant friction. Establishing clear, mutually agreed-upon data sharing agreements, technical mech-

Table 13.1 Key foundational research gaps hindering trustworthy AI/ML-enhanced supply chain digital twins

Foundational gap area	Specific challenges/needs for AI/ML-driven supply chain DTs
Verification, validation, and uncertainty quantification (VVUQ)	Need for robust, *continual* VVUQ methods for dynamic, adaptive, AI/ML-infused systems. Major gaps exist in VVUQ for extrapolation scenarios, rare event prediction (critical for supply chain risks), and quantifying confidence in AI/ML model outputs. Lack of reporting standards hinders trust
Hybrid modeling (mechanistic modeling + AI/ML)	Lack of mature theory and practical methods for effectively combining physics-based/domain-knowledge models with data-driven AI/ML. Need for integration techniques, validation strategies, and VVUQ specific to hybrid models to ensure robustness and interpretability in supply chain decisions
System integration and scalability	Challenges in reliably integrating component/subsystem DTs across the complex supply chain system of systems, especially with varying model fidelities. Need for scalable algorithms (AI/ML, optimization, VVUQ) and computational infrastructure to handle large-scale supply chain problems in near real time
Data quality, integration and standards	Difficulties in integrating heterogeneous, multimodal supply chain data streams due to lack of standards for data formats, metadata, and semantics. Need for robust methods to handle data quality issues (noise, sparsity, bias) and ensure data provenance, especially as input for sensitive AI/ML models
Autonomous decision-making under uncertainty	Need for scalable optimization methods that incorporate sophisticated risk metrics and handle extreme events common in supply chain. Gaps in seamlessly integrating AI/ML techniques (e.g., reinforcement learning) with data assimilation and control theory for dynamic supply chain optimization. Need for AI-driven optimal sensor steering/OED relevant to supply chain decisions
Ethics, governance, and security	Unresolved challenges in data privacy, ownership, and liability across multi-enterprise SCDTs. Heightened cybersecurity risks in interconnected cyber-physical supply chain systems. Potential for AI bias amplification. Lack of specific governance frameworks for trustworthy deployment and accountability
Sustainability and lifecycle management	Need to establish frameworks for managing SCDTs as evolving digital assets throughout their lifecycle, ensuring long-term reliability, maintainability, and trustworthiness beyond initial deployment

anisms for privacy-preserving data analysis, and robust data ownership models are essential prerequisites for the multi-enterprise data flows required by comprehensive SCDTs.

Security Vulnerabilities in Cyber-Physical Supply Chain Systems The tight integration between the physical supply chain and its virtual representation, central to the digital twin concept, significantly expands the cybersecurity attack surface. SCDTs are cyber-physical systems where vulnerabilities can have direct real-world consequences. Malicious actors could exploit these vulnerabilities in numerous ways: intercepting data feeds to gain competitive intelligence about shipments or production; injecting false sensor data (e.g., incorrect temperatures for cold chain logistics, fake vibration readings for machinery) to trigger erroneous alerts, unnecessary costly maintenance, or mask actual problems; manipulating the SCDT's control outputs to disrupt physical processes (e.g., misdirecting autonomous vehicles, altering warehouse sorting logic, interfering with production schedules); or using a compromised partner's connection to the shared SCDT ecosystem as an entry point to attack other participants. Securing this complex environment requires a holistic cybersecurity strategy extending beyond traditional IT security to encompass OT, IoT devices, data links between partners, the SCDT platform itself, and the AI/ML algorithms within it, demanding robust measures for authentication, encryption, data integrity validation, anomaly detection, and rapid incident response across the entire interconnected supply chain network.

Potential for Bias Amplification by AI in Supply Chain Decisions AI and ML algorithms learn from data, and if the training data reflects existing societal or historical biases, or if biases are introduced during model development, the resulting AI models can perpetuate or even amplify unfair outcomes. Models may yield discriminatory results due to biases. In supply chain, this could manifest in various ways: AI-driven routing algorithms optimizing solely for speed or cost might disproportionately route heavy truck traffic through disadvantaged communities already suffering from pollution and noise, neglecting equity considerations. Sourcing algorithms assessing supplier risk based on historical data or potentially biased news sentiment analysis might unfairly penalize suppliers from certain geographic regions or demographic groups, reducing supply base diversity and resilience. Demand forecasting or inventory allocation models might systematically underestimate the needs of underserved populations if trained primarily on historical sales data that reflects limited access rather than true underlying demand. Without careful design, bias auditing, and ongoing monitoring, AI within SCDTs could inadvertently lead to inequitable resource distribution, unfair competitive practices, and reputational damage.

Liability and Accountability in Case of SCDT-Informed Failures As SCDTs, potentially enhanced with AI/ML for autonomous decision-making, take on more critical roles in operational management, questions of liability become paramount. It is important to establish guidelines for identifying liability. If an autonomous SCDT makes a decision based on a faulty AI prediction or flawed data—leading to a major delivery failure, a production shutdown, or even a safety incident—determining who is legally and financially responsible becomes exceptionally complex. Is it the developer of the SCDT platform, the creator of the specific AI algorithm, the provider(s) of the underlying data, the organization that deployed the SCDT, or the human operator who may (or may not) have overseen the autonomous action? The distributed nature of supply chain, the involvement of multiple commercial entities, and the often opaque nature of complex AI models ("black box" problem) make tracing causality and assigning liability a significant legal and organizational challenge, potentially hindering the adoption of more advanced autonomous capabilities.

The Need for Comprehensive Governance Frameworks Addressing these interwoven ethical, security, privacy, and liability concerns requires more than ad hoc solutions; it necessitates the development and adoption of robust governance frameworks specifically designed for AI/ML-enhanced SCDTs. These frameworks must move beyond existing regulations and standards, providing clear guidelines and protocols for: multiparty data sharing, usage rights, privacy protection, and security standards; AI model transparency, explainability, bias detection and mitigation procedures; cybersecurity best practices for interconnected cyber-physical supply chain systems; and principles for establishing accountability and liability in complex, multiagent environments. Proactive development of such governance structures, ideally through collaborative efforts involving industry stakeholders, regulators, and researchers, is essential for building the necessary trust and ensuring the responsible and sustainable deployment of these powerful technologies.

13.3.3 Future Directions: Toward Trustworthy and Sustainable Supply Chain Digital Twins

Moving AI/ML-enhanced SCDTs from a promising concept to a widespread, reliable, and value-generating reality for supply chain requires coordinated efforts across research, development, policy, and education, centered on building systems that are both trustworthy and sustainable over their operational lifetime. Key future directions include:

Executing an Integrated Research Agenda A fundamental shift is required away from siloed research efforts toward a holistic, integrated research agenda. This means tackling the interdependent challenges of SCDTs—spanning virtual representation (including AI/ML modeling), physical counterpart data, bidirectional feedback loops (including DA and control), VVUQ, ethics, and security—as a cohesive whole, rather than as isolated problems. There is a need for research on new cross-cutting programs focused on advancing the mathematical, statistical, and computational foundations underpinning SCDTs. Connecting theoretical advancements directly to real-world supply chain problems (e.g., resilience optimization, last-mile delivery efficiency, sustainable sourcing) will ensure relevance, guide research priorities, and accelerate the translation of innovations into impactful solutions.

Developing Sustainable Digital Twins with Lifecycle Management The long-term viability of SCDTs depends on treating them not as one-off projects, but as enduring digital assets that require ongoing investment and management. This necessitates establishing robust frameworks and practices for SCDT lifecycle management. Future efforts must focus on developing methodologies and tools for systematically maintaining, updating, and evolving all components of the SCDT—including the AI/ML models, underlying software platforms, data integration pipelines, sensor network configurations, and virtual representations—as the physical supply chain and operational requirements change over time.

Fostering Cross-Disciplinary Research, Translation, and Collaboration Breaking down existing silos between disciplines and sectors is also important. As the foundations of SCDTs mature, it becomes crucial to actively foster cross-disciplinary learning and identify common research themes and solutions that can be translated across different application domains. This involves creating venues (like workshops) for researchers and practitioners from supply chain, computer science, statistics, operations research, engineering, and other fields to share insights and best practices related to AI/ML integration, VVUQ, data handling, etc. In addition, bridging the gap between academic research and industrial application requires strengthening collaboration between universities, government agencies, and industry partners specifically involved in supply chain. Given the global nature of modern supply chains, international collaboration is also essential. Establishing forums and mechanisms for the collaborative exchange of data and models (where appropriate and secure, considering privacy and IP) can significantly accelerate progress.

The multifaceted path forward requires a holistic and interconnected strategy. Figure 13.2 visually represents this required ecosystem approach for advancing trustworthy and sustainable AI/ML-enhanced SCDTs. The figure centers on this ultimate goal and illustrates the important, interdependent elements needed to achieve it. These elements include establishing an integrated foundational research agenda, fostering robust cross-disciplinary and cross-sector collaboration (including international partnerships), developing essential data and model sharing frameworks with appropriate standards, implementing lifecycle management practices with a focus on sustainability and embedding strong governance and ethical guidelines. The figure underscores that progress is not contingent on advancing any single

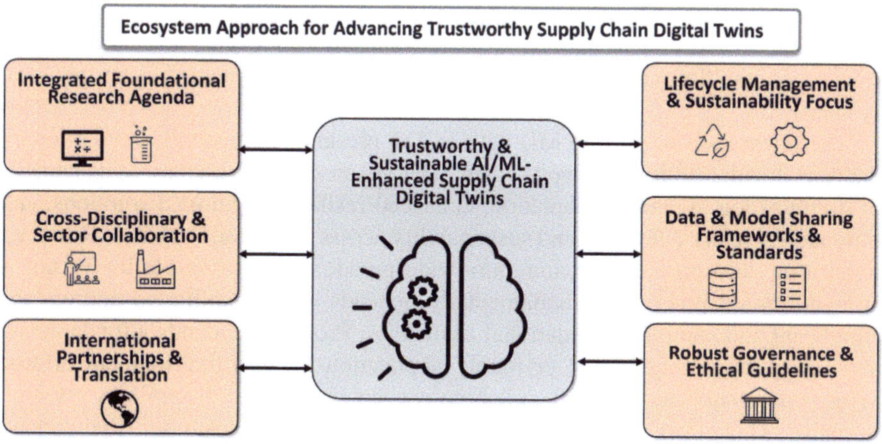

Fig. 13.2 Ecosystem approach for advancing trustworthy SCDTs

element in isolation, but rather on nurturing the synergistic development of this entire interconnected ecosystem.

13.4 Conclusion

In this chapter, we have explored the synergistic integration of AI and ML with SCDT, highlighting its transformative potential for supply chain management and optimization. This combination promises to move beyond traditional supply chain methods by enabling unprecedented levels of real-time visibility, sophisticated predictive analytics for demand and disruptions, data-driven predictive maintenance strategies for critical assets, and even pathways toward autonomous decision-making across complex logistics, inventory, and sourcing operations. The overarching prospect is the creation of significantly more efficient, agile, responsive, and resilient supply chains capable of navigating the volatilities of the modern global landscape.

However, a crucial insight central to our analysis is that realizing this significant potential is critically contingent upon confronting and overcoming substantial foundational research challenges. The widespread deployment of reliable, trustworthy, and scalable AI/ML-enhanced SCDTs is currently hindered by significant gaps. These research needs span fundamental mathematical domains, such as developing scalable algorithms for optimization under deep uncertainty and establishing theory for complex hybrid models. They encompass critical statistical challenges, including the creation of robust methods for VVUQ applicable to dynamic, adaptive AI/ML systems operating with imperfect data, especially during extrapolation or for rare events. Significant computational hurdles exist related to integrating heterogeneous systems, ensuring algorithm scalability for real-time performance, and managing vast, diverse data streams securely and effectively. Furthermore, crucial translational gaps persist concerning the development of effective human-digital twin interaction paradigms, robust ethical and governance frameworks for data sharing and decision accountability, and ensuring cybersecurity in interconnected cyber-physical systems.

The integration of AI and ML with SCDT presents a potentially revolutionary pathway for the future of supply chain. The vision encompasses systems capable of optimization, dynamic adaptation, enhanced resilience against disruptions, and ultimately, greater efficiency and sustainability across global value chains. However, translating this powerful vision into widespread and reliable reality is not a straightforward process. It fundamentally depends on our ability to address the significant and complex foundational challenges. Progress demands a fundamental shift away from fragmented or purely application-focused development efforts. What is required is a concerted, collaborative, and deeply interdisciplinary research initiative. This effort must bring together diverse expertise spanning supply chain domain knowledge, computer science (including AI/ML), mathematical and statistical sciences, systems engineering, cybersecurity, ethics, and social sciences, fostering partnerships between academia, industry, and government entities. The central pillars of this essential research effort must be squarely focused on:

- **Building Trust:** Developing transparent, interpretable, and ethically grounded systems whose operations and limitations can be understood and relied upon by diverse stakeholders across the supply chain.
- **Ensuring Reliability:** Placing robust VVVUQ at the center of development and deployment. This involves creating and applying rigorous methodologies to quantify confidence, understand failure modes, validate performance under real-world conditions (including extrapolation and rare events), and ensure the stability and correctness of dynamic, adaptive AI/ML-infused SCDTs.
- **Managing Complexity:** Developing theoretical frameworks, scalable algorithms, and practical methodologies capable of handling the inherent complexities such as multiscale interactions, heterogeneous and imperfect data streams, seamless system integration across organizational boundaries, dynamic adaptation, and intricate cyber-physical security vulnerabilities.

The future promise of truly optimized, adaptive, and resilient supply chains—capable of thriving amidst global uncertainty and meeting the increasing demands for sustainability and responsiveness—is inextricably tied to the success of tackling these foundational challenges. Investing purposefully in this fundamental research, grounded in real-world supply chain needs, is not merely an academic or technological exercise; it represents a critical strategic imperative for building the intelligent, reliable, and trustworthy supply chain ecosystems required for the future.

References

Abd Wahab, N.H., Hasikin, K., Lai, K.W., Xia, K., Bei, L., Huang, K., & Wu, X. (2024). Systematic review of predictive maintenance and digital twin technologies challenges, opportunities, and best practices. *PeerJ Computer Science, 10*, e1943.

Adewusi, A.O., Komolafe, A.M., Ejairu, E., Aderotoye, I.A., Abiona, O.O., & Oyeniran, O.C. (2024). The role of predictive analytics in optimizing supply chain resilience: A review of techniques and case studies. *International Journal of Management & Entrepreneurship Research, 6*(3), 815–837.

Aljohani, A. (2023). Predictive analytics and machine learning for real-time supply chain risk mitigation and agility. *Sustainability, 15*(20), 15088.

Alshathri, S., Hemdan, E.E.D., El-Shafai, W., & Sayed, A. (2023). Digital twin-based automated fault diagnosis in industrial IoT applications. *Computers, Materials & Continua, 75*(1), 183–196

Anis, M.D., Taghipour, S., & Lee, C.G. (2020). Optimal rule estimation: A state-of-art digital twin application. In *2020 Annual Reliability and Maintainability Symposium (RAMS)* (pp. 1–7). IEEE.

Baryannis, G., Dani, S., & Antoniou, G. (2019). Predicting supply chain risks using machine learning: The trade-off between performance and interpretability. *Future Generation Computer Systems, 101*, 993–1004.

Biller, B., Yi, J., & Biller, S. (2023). A practitioner's guide to digital twin development. In *Tutorials in Operations Research: Advancing the Frontiers of OR/MS: From Methodologies to Applications, INFORMS* (pp. 198–227).

Chen, I.J., & Kitsis, A.M. (2017). A research framework of sustainable supply chain management: The role of relational capabilities in driving performance. *The International Journal of Logistics Management, 28*(4), 1454–1478.

Committee on Mathematical Foundations of Verification, Validation, and Uncertainty Quantification National Academies of Sciences, Engineering, and Medicine. (2012). *Assessing the reliability of complex models: mathematical and statistical foundations of verification, validation, and uncertainty quantification*. National Academies Press.

Ernst, D. (2002). Global production networks and the changing geography of innovation systems. implications for developing countries. *Economics of Innovation and New Technology, 11*(6), 497–523.

He, B., Liu, L., & Zhang, D. (2021). Digital twin-driven remaining useful life prediction for gear performance degradation: A review. *Journal of Computing and Information Science in Engineering, 21*(3), 030801.

Hegazy Abdelghany Mohamed Ammar, M. (2021). Advanced digital twins for conditions monitoring, examinations, diagnosis and predictive remaining lifecycles based artificial intelligence. PhD thesis, Brunel University London.

Holloway, S. (2025). *Navigating supply chain challenges with digital transformation for agility and resilience*. Preprints. https://doi.org/10.20944/preprints202502.0970.v1

Huang, H., Yang, L., Wang, Y., Xu, X., & Lu, Y. (2021). Digital twin-driven online anomaly detection for an automation system based on edge intelligence. *Journal of Manufacturing Systems, 59*, 138–150.

Hugos, M.H. (2024). *Essentials of supply chain management*. John Wiley & Sons.

Khan, N.K. (2024). Designing of digital supply chain twins to improve supply chain resilience using machine learning and simulation models. PhD thesis, State University of New York at Binghamton.

Khedr, A.M., et al. (2024). Enhancing supply chain management with deep learning and machine learning techniques: A review. *Journal of Open Innovation: Technology, Market, and Complexity*, 100379.

Kleindorfer, P.R., & Saad, G.H. (2005). Managing disruption risks in supply chains. *Production and Operations Management, 14*(1), 53–68.

Kobayashi, K., & Alam, S.B. (2024). Explainable, interpretable, and trustworthy AI for an intelligent digital twin: A case study on remaining useful life. *Engineering Applications of Artificial Intelligence, 129*, 107620.

Lechler, S., Canzaniello, A., Roßmann, B., von der Gracht, H.A., & Hartmann, E. (2019). Real-time data processing in supply chain management: revealing the uncertainty dilemma. *International Journal of Physical Distribution & Logistics Management, 49*(10), 1003–1019.

Long, Y., Kroeger, S., Zaeh, M.F., & Brintrup, A. (2025). Leveraging synthetic data to tackle machine learning challenges in supply chains: challenges, methods, applications, and research opportunities. *International Journal of Production Research*, 1–22.

Lou, P., Zhong, Y., Hu, J., Fan, C., & Chen, X. (2023). Digital-twin-driven AGV scheduling and routing in automated container terminals. *Mathematics, 11*(12), 2678.

Maheshwari, P., Kamble, S., Belhadi, A., Venkatesh, M., & Abedin, M.Z. (2023). Digital twin-driven real-time planning, monitoring, and controlling in food supply chains. *Technological Forecasting and Social Change, 195*, 122799.

Manners-Bell, J. (2023). *Supply chain risk management: How to design and manage resilient supply chains*. Kogan Page Publishers.

Mihai, S., Yaqoob, M., Hung, D.V., Davis, W., Towakel, P., Raza, M., Karamanoglu, M., Barn, B., Shetve, D., Prasad, R.V., et al. (2022). Digital twins: A survey on enabling technologies, challenges, trends and future prospects. *IEEE Communications Surveys & Tutorials 24*(4), 2255–2291.

Muthukalyani, A.R. (2024). Enhancing supply chain agility with real-time data analytics. *International Journal of Data Science and Analytics (IJDSA), 2*(1), 1–9.
NASEM. (2024). *National Academy of Engineering and National Academies of Sciences, Engineering, and Medicine: Foundational research gaps and future directions for digital twins*. The National Academies Press. https://doi.org/10.17226/26894. https://nap.nationalacademies.org/catalog/26894/foundational-research-gaps-and-future-directions-for-digital-twins
Neway, A.S. (2024). *Digital supply chain: Navigating the future operation*. Abegaz Sahilu Neway.
Rathore, M.M., Shah, S.A., Shukla, D., Bentafat, E., & Bakiras, S. (2021). The role of AI, machine learning, and big data in digital twinning: A systematic literature review, challenges, and opportunities. *IEEE Access, 9*, 32030–32052.
Sarimveis, H., Patrinos, P., Tarantilis, C.D., & Kiranoudis, C.T. (2008). Dynamic modeling and control of supply chain systems: A review. *Computers & Operations Research, 35*(11), 3530–3561.
Sheffi, Y. (2015). *The power of resilience: How the best companies manage the unexpected*. MIT Press.
Singh, P. (2023). Digital transformation in supply chain management: Artificial intelligence (AI) and machine learning (ML) as catalysts for value creation. *International Journal of Supply Chain Management, 12*(6), 57–63.
Soykan, B., & Rabadi, G. (2024). Optimizing humanitarian logistics with deep reinforcement learning and digital twins. In *2024 Annual Modeling and Simulation Conference (ANNSIM)* (pp. 1–13). https://doi.org/10.23919/ANNSIM61499.2024.10732882
Soykan, B., Blanc, G., & Rabadi, G. (2025). A proof-of-concept digital twin for real-time simulation: Leveraging a model-based systems engineering approach. *IEEE Access*, 1–1. https://doi.org/10.1109/ACCESS.2025.3557367
Spekman, R.E., Kamauff Jr. J.W., & Myhr, N. (1998). An empirical investigation into supply chain management: a perspective on partnerships. *Supply Chain Management: An International Journal, 3*(2), 53–67.
Themistocleous, M., Irani, Z., & Love, P.E. (2004). Evaluating the integration of supply chain information systems: A case study. *European Journal of Operational Research, 159*(2), 393–405.
Xu, Q., Ali, S., & Yue, T. (2021). Digital twin-based anomaly detection in cyber-physical systems. In *2021 14th IEEE Conference on Software Testing, Verification and Validation (ICST)* (pp. 205–216). IEEE.
Zhang, L., Yang, C., Yan, Y., Cai, Z., & Hu, Y. (2024). Automated guided vehicle dispatching and routing integration via digital twin with deep reinforcement learning. *Journal of Manufacturing Systems, 72*, 492–503.

MIX
Papier aus verantwortungsvollen Quellen
Paper from responsible sources
FSC® C105338

FSC
www.fsc.org

If you have any concerns about our products,
you can contact us on
ProductSafety@springernature.com

In case Publisher is established outside the EU,
the EU authorized representative is:
Springer Nature Customer Service Center GmbH
Europaplatz 3, 69115 Heidelberg, Germany

Printed by Libri Plureos GmbH
in Hamburg, Germany